from
THE
QUARTERLY
REVIEW
OF
BIOLOGY

44135   #45.00

8/20/20

Changing Connectomes

Changing Connectomes

Evolution, Development, and Dynamics in Network Neuroscience

Marcus Kaiser

The MIT Press
Cambridge, Massachusetts
London, England

© 2020 Massachusetts Institute of Technology

All rights reserved. No part of this book may be reproduced in any form by any electronic or mechanical means (including photocopying, recording, or information storage and retrieval) without permission in writing from the publisher.

This book was set in ITC Stone Serif Std and ITC Stone Sans Std by New Best-set Typesetters Ltd. Printed and bound in the United States of America.

Library of Congress Cataloging-in-Publication Data

Names: Kaiser, Marcus, author.
Title: Changing connectomes : evolution, development, and adaptations in network neuroscience / Marcus Kaiser.
Description: Cambridge, Massachusetts : The MIT Press, [2020] | Includes bibliographical references and index.
Identifiers: LCCN 2019059139 | ISBN 9780262044615 (hardcover)
Subjects: LCSH: Neural networks (Neurobiology) | Developmental neurobiology. | Nervous system—Evolution. | Brain—Growth. | Brain—Diseases. | Brain—Degeneration.
Classification: LCC QP363.3 .K34 2020 | DDC 612.6/4018—dc23
LC record available at https://lccn.loc.gov/2019059139

10 9 8 7 6 5 4 3 2 1

For Eileen and Felicitas

Contents

Preface ix

1 Introduction 1

Part I Connectome Structure 5
2 Features of Complex Networks 7
3 Evolution of Neural Systems 37
4 Organization of Neural Systems 43

Part II Connectome Maturation 65
5 Brain Development 67
6 Layer Formation 79
7 Axonal Growth 89
8 Formation of Hubs 107
9 Module Formation 119
10 Cortical Folding 131

Part III Connectome Changes 141
11 Development and Aging 143
12 Neurodevelopmental Disorders 153
13 Neurodegenerative Disorders 165
14 Recovery from Injury 177
15 Brain Stimulation Effects 187

Glossary 199
References 205
Index 245

Preface

The human brain undergoes massive changes during brain development, ranging from early childhood and adolescence to old age. The field of network neuroscience has provided snapshots at different stages, and, through data analysis and computational modeling, mechanisms leading to changing connectomes begin to emerge. This book aims to provide an overview of network features of the brain, how these features emerge during evolution and brain development, and how alterations ranging from normal aging to developmental and neurodegenerative disorders affect brain networks. The book will give biomedical researchers, who are often only aware of neuroimaging research in humans, information about the knowledge on connectome development in other species. By including an introduction to concepts of network analysis, it will also be accessible to researchers who are new to the field of connectomics. Finally, through Matlab/Octave code examples (available at https://mitpress.mit.edu/changing-connectomes), it will allow computational neuroscience researchers to understand and extend the shown mechanisms of connectome development.

I took my first step into connectomics in the summer of the year 2000, when I did a research internship, while still an undergraduate at Ruhr University Bochum, with Malcolm Young at Newcastle University. In these early days, only connectivity from animal models was available, ranging from the roundworm *Caenorhabditis elegans* to rat, cat, and rhesus monkey. When I worked with Claus Hilgetag during my PhD studies from 2002 to 2005, knowing about how only a single human brain would be connected seemed like a distant dream. Nowadays, following pioneering studies such as the Human Connectome Project in the United States, large data sets of human brain networks in health and disease are available. Within Europe, the UK Biobank Imaging Study is currently collecting structural and functional brain connectivity data of 100,000 subjects, aiming to complete data collection in 2022. Finally, longitudinal studies of brain development before birth, in early childhood, and during the life span are being performed.

Writing a book related to the fastest growing area of brain research, network neuroscience, is challenging. In particular, given that writing books happens along with other academic commitments, you might ask, *"I wonder who they found to pull that off?"* I would therefore like to thank the School of Computing at Newcastle University for granting me one semester of research leave to work on this book. I am also indebted to my family, in particular during the final months of finishing this book.

For a first step toward neuroscience and connectome research, I would like to thank Klaus-Peter Hoffmann and Malcolm Young, respectively. Much of this work relies on fruitful discussions and collaborations with my colleagues, mainly Claus Hilgetag, but also Rolf Kötter, Olaf Sporns, Herbert Jaeger, Miles Whittington, and Arjen van Ooyen. I also benefited from discussions with current and former members of my research team. For providing feedback on parts of this book, I would like to thank Alexandros Goulas, Ann-Shyn Chiang, Arjen van Ooyen, Bruno Mota, Cheol Han, Cornelis Stam, David Willshaw, Georg Striedter, Hongkui Zeng, John-Paul Taylor, Larry Swanson, Lisa Ronan, Markus Butz-Ostendorf, Michael Nitsche, Olaf Sporns, Petra Vertes, Roman Bauer, Sven Bestmann, Thiebaut de Schotten, and Yujiang Wang.

Finally, I would also like to thank my editor at the MIT Press, Bob Prior, for his support in getting this book project off the ground and landing safely now.

1 Introduction

How does the brain change over time? While researchers are beginning to understand more about how the set of connections between neurons and regions, the connectome, is organized, this book describes how such an organization arises during development and evolution, how it changes in health and disease, and how external interventions can alter its architecture. For this aim, we describe mechanistic insights, based on computational models and experimental studies, that can link the changes at the local level of individual neurons to the observed large-scale alterations in connectivity between brain regions.

Over the past several years, connectome information about large cohorts of human subjects and patients at different stages of brain development or disease progression has become available. There are continuing efforts to increase data quality, to enable data sharing, and to analyze brain network architecture. There is increasing availability of longitudinal data, measuring the same subjects at multiple time points. Along with connectome data of the early stages, before and shortly after birth, this allows us for the first time to observe the development of connectomes.

Given the availability of both data on brain networks and tools to describe the organization and behavior of these networks, the field of network neuroscience can inform how we look at connectome changes. There are different approaches that one can take to analyze connectome changes. First, one may look at changes in the network organization at different time points using network science approaches. This analysis of network features gives a first insight into which network functions—for example, integration and segregation of information flows—might be altered. Second, one may simulate neural activity within the network to understand how changes in network structure influence network behavior. Finally, one may use computational models to evaluate the mechanisms that lead to the observed changes in network structure. Such approaches can help us to understand how connectome changes arise and change brain function, and they could suggest hypotheses that can be tested experimentally in the future.

There are also more direct applications of network neuroscience in understanding connectome changes. As will be described, there are distinct network changes for brain disorders. Computer models will be essential to inform diagnosis and treatment of individual patients. As each case is different, it is impossible to have an experimental animal or clinical human study with exactly the same condition as found in an individual. In addition, given dozens of variables that play a role, and with interactions between variables influencing each other, it will be impossible to provide mathematical equations that describe the relationships between all variables. The only solution for better diagnosis and treatment of individual patients is a personalized computational representation, based on connectome, physiome, genome, and other available data. With such a system, the plausibility of different disease origins and the outcomes of different interventions can be tested in silico to find the most suitable option for an individual patient.

The study of mechanisms of network changes can also be helpful for designing experimental studies, for understanding the link between brain network structure and performance, and for improving the design and development of artificial neural networks. The range of applications that can benefit from an understanding of connectome changes includes, for example, the following:

- The design of artificial neural networks—for example, for deep learning—includes features of connectomes such as layers but only includes a small subset of mechanisms for network growth and development. This particularly limits the ability of such systems to provide more complex behavior such as multimodal integration, adaptivity to new environments, and learning from small training data sets. Building networks that grow based on mechanisms identified in biological neural networks might provide new breakthroughs in this field.
- Knowing about the link between connectome structure and function will help us to understand why network features arise during individual brain development and during the evolution of neural systems. Not all network features might have a direct or strong link to cognitive processing.
- Knowing the developmental origin of brain diseases provides another biomarker that can be used to help with diagnosis, the stratification of the patient cohort, and treatment planning. Understanding the developmental pathways of network changes could, given the current brain network of a patient, help to predict which factors played a role in the genesis of these connectome changes.

Understanding the mechanisms that lead to network changes can help to improve brain function following brain injury by facilitating the design of rehabilitation interventions that increase positive network changes while trying to prevent network changes that have a negative effect on brain function.

- For brain stimulation, while there are some models about immediate stimulation effects, during and shortly after stimulation, it will be crucial to understand long-term effects in order to predict effects and minimize negative side effects.

In part I, before we can look at connectome changes, we first need to describe how brain networks can be measured and how we can analyze their features. Chapter 2 provides an overview of network reconstruction and of the analysis of topological and spatial features. Furthermore, I show how activity in these networks can be modeled by giving a brief overview of dynamic features of brain networks. Chapter 3 shows how topological features arise during brain evolution, starting with simple nerve nets and moving on to modular and hierarchical networks. Chapter 4 gives an overview of the architecture of brain networks for the organisms for which we already have full or partial information about their connectomes: *Caenorhabditis elegans*, fruit fly, pigeon, mouse, rat, ferret, cat, rhesus monkey, marmoset monkey, and human. Part II discusses the maturation of network features during individual brain development. Chapter 5 shows how regional patterns such as cortical maps can be formed and how genetic factors, competition, and homeostasis can induce these patterns. Chapter 6 shows how layers can form; it includes experimental and computational results indicating the roles of cell growth, cell migration, and cell death. Chapter 7 looks at axon growth and the formation of synaptic connections determining principles of initial connection establishment. Chapter 8 looks at network hubs, outlining different hub types and different mechanisms that can generate hubs during brain development. Chapter 9 describes how modules, enabling segregated information processing, can arise due to developmental time windows and genetic factors. Chapter 10 describes how cortical folding changes during development and what principles and mechanisms might cause these changes. Part III looks at how connectomes change during the life span in health and disease and how interventions can interfere with these processes. Chapter 11 is about healthy brain development, outlining changes until adulthood. Chapter 12 talks about changes due to neurodevelopmental disorders such as schizophrenia, autism spectrum disorders, major depression, epilepsy, and Tourette's syndrome as well as about underlying mechanisms for these changes. Chapter 13 deals with age-related disorders such as Alzheimer's disease, Lewy body dementia, and Parkinson's disease as well as with models of disease progression. Chapter 14 describes how connectomes react to lesions, caused by stroke, traumatic brain injury, or loss of peripheral input, and how computational models can be used to test underlying mechanisms for these changes. Chapter 15 highlights the emerging role of brain stimulation, being used on patients but also on healthy subjects, in altering the dynamics but also the topology of brain networks.

How can we determine brain connectivity and how can we analyze brain networks? The next chapter will show how structural and functional connectivity can be determined and how topological, spatial, and dynamic characteristics of the network can be analyzed.

1 Connectome Structure

The connectome architecture can be studied in several ways. It is important to keep in mind that scope, resolution, and parcellation of the nervous system influence the observed network features. The scope of the network could range from the cortex and subcortical structures to a network that also includes the peripheral nervous system. Therefore, when comparing between different studies in humans or between different species, some changes in the network are already due to such differences.

While we have a complete network of the nervous system for *C. elegans*, studies of the human brain are usually limited to the cortex and some subcortical structures. Even for the central nervous system in humans, including the cerebellum, which has 10 times as many neurons as the telencephalon (Herculano-Houzel, 2009), leads to a completely different network. The resolution could be at the level of fiber tract connectivity between regions, neuronal connectivity between smaller patches of brain tissue such as cortical columns, or axonal connectivity between individual neurons.

With higher resolution, the number of network nodes increases drastically: at the regional resolution a human cortical network might consist of 100 regions, whereas at the resolution of individual neurons it consists of 10 billion nodes. Correspondingly, the *edge density*, the proportion of existing connections relative to all possible connections, changes from 10% at the regional resolution to 0.0001% at the neuronal resolution.

Finally, even within a resolution level, parcellation schemes lead to a wide range of network nodes. For the human brain, regional parcellations of the cortex vary from 68 regions (Hagmann et al., 2008) to 360 regions (Glasser et al., 2016) with a wide range of different anatomical, functional, or multimodal parcellation approaches (Eickhoff, Constable, and Yeo, 2018; Eickhoff, Yeo, and Genon, 2018).

2 Features of Complex Networks

The set of connections in neural systems, now called the connectome (Sporns et al., 2005), has been the focus of neuroanatomy for more than a hundred years (Ramón y Cajal, 1892; His, 1888). However, it has attracted recent interest due to the increasing availability of network information at the global (Felleman and Van Essen, 1991; Scannell et al., 1995; Burns and Young, 2000; Tuch et al., 2003) and local levels (White et al., 1986; Denk and Horstmann, 2004; Lichtman et al., 2008; Seung, 2009) as well as the availability of network analysis tools that can elucidate the link between structure and function of neural systems. Within the neuroanatomical network (structural connectivity), the nonlinear dynamics of neurons and neuronal populations result in patterns of statistical dependencies (functional connectivity) and causal interactions (effective connectivity), defining three major modalities of complex neural systems (Sporns et al., 2004). How is the network structure related to its function, and what effect does changing network components have (Kaiser, 2007)? Since 1992 (Achacoso and Yamamoto, 1992; Young, 1992), tools from network analysis (Costa, Rodrigues, et al., 2007) have been applied to study these questions in neural systems.

What are the benefits of using network analysis in neuroimaging research? First, networks provide an abstraction that can reduce the complexity when dealing with neural networks. Human brains show a large variability in size and surface shape (Van Essen and Drury, 1997). Network analysis, by hiding these features, can help to identify similarities and differences in the organization of neural networks. Second, the overall organization of brain networks has been proven reliable in that features such as small worldness and modularity, present but varying to some degree, are found in all human brain networks (and those of other species, too). Third, using the same frame of reference, given by the identity of network nodes as representing brain regions, both comparisons between subjects as well as comparisons of different kinds of networks (e.g., structural vs. functional) are feasible (Rubinov and Sporns, 2010).

The analysis of networks originated from the mathematical field of graph theory (Diestel, 1997), later leading to percolation theory (Stauffer and Aharony, 2003) or social network analysis (Wasserman and Faust, 1994). In 1736, Leonhard Euler worked on the problem of crossing all bridges over the river Pregel in Königsberg (now Kaliningrad) exactly once and returning to the origin, a path now called an Euler tour. These and other problems can be studied by using graph representations. Graphs are sets of nodes and edges. Edges can either be undirected, going in both directions, or directed (arcs or arrows) in that one can go from one node to the other but not in the reverse direction. A *path* is a walk through the graph where each node is only visited once. A *cycle* is a closed walk, meaning a path that returns back to the first node. A graph could also contain *loops* that are edges that connect a node to itself; however, for analysis purposes we only observe simple graphs without loops. In engineering, graphs are called networks if there is a source and sink of flow in the system and a capacity for flow through each edge (e.g., flow of water or electricity). However, following conventions in the field of network science, I will denote all brain connectivity graphs as networks.

For brain networks, nodes could be neurons or cortical areas and edges could be axons or fiber tracts. Thus, edges could refer to the *structural connectivity* of a neural network. Alternatively, edges could signify correlations between the activity patterns of nodes forming *functional connectivity*. Finally, a directed edge between two nodes could exist if activity in one node modulates activity in the other node, forming *effective connectivity* (Sporns et al., 2004). Network representations are an abstract way to look at neural systems. Among the factors missing from network models of nodes, say brain areas, are the location, the size, and the functional properties of the nodes. In contrast, geographical or spatial networks also give information about the spatial location of a node. Two- or three-dimensional Cartesian coordinates in a metric space indicate the location of neurons or areas.

Network analysis techniques can be applied to the analysis of brain connectivity, and I will discuss structural connectivity as an example. Using neuroanatomical or neuroimaging techniques, one can test which nodes of a network are connected, that is, whether projections in one or both directions exist between a pair of nodes (see figure 2.1A). How can information about brain connectivity be represented? If a projection between two nodes is found, the value one is entered in the adjacency matrix; the value zero defines absent connections or cases where the existence of connections was not tested (see figure 2.1B). The memory demands for storing the matrix can be prohibitive for large networks as N^2 elements are stored for a network of N nodes. As most neuronal networks are sparse, storing only information about existing edges can save storage space. Using a list of edges, the adjacency list (see figure 2.1C) stores each edge in one

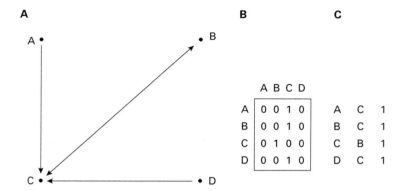

Figure 2.1
Representations of networks. (A) Directed graph with two directed edges or arcs (AC and DC) and one undirected edge being equivalent to a pair of directed edges in both directions (BC). (B) The same graph can be represented in a computer using an adjacency matrix where a value of 1 denotes the existence of an edge and 0 the absence of an edge. In this example, rows show outgoing connections of a node and columns show incoming connections. (C) Sparse matrices (few edges) can also be represented as adjacency lists to save memory. Each edge is represented by the source node, the target node, and the weight of the edge (here: uniform value of 1).

row listing the source node, the target node, and—for networks with variable connection weight—the strength of a connection.

2.1 Reconstructing Connectomes

How can one get information about brain connectivity between regions, the macroscopic connectome (Akil et al., 2011)? The classical way to find out about structural connectivity is to inject dyes into a brain region. The dye is then taken up by dendrites and cell bodies and travels within a neuron either in an anterograde (from soma to synapse) or a retrograde (from synapse to soma) direction. Typical dyes are horseradish peroxidase, fluorescent microspheres, Phaseolus vulgaris-leucoagglutinin, Fluoro-Gold, Cholera toxin subunit B, DiI, and tritiated amino acids. Allowing some time for the tracers to travel, which could be several weeks for the large human brain, the neural tissue can be sliced up, and dyes can indicate the origin and target of cortical fiber tracts. Whereas this approach yields high-resolution information about structural connectivity, it is an invasive technique usually unsuitable for human subjects (however, there are some postmortem studies). In the following, I will therefore present noninvasive neuroimaging solutions to yield structural and functional connectivity.

The workflow for yielding connectivity data starts with anatomical magnetic resonance imaging (MRI) scans with high resolution (see figure 2.2). These scans are later used to register the location of brain regions. For establishing functional connectivity, a time series of brain activity in different voxels or regions can be derived. The correlation between the time series of different voxels or, using aggregated measures, brain regions can be detected and represented as a correlation matrix (with values ranging from –1 to 1). This matrix either can directly be interpreted as a weighted network or can be transformed into a binary matrix in that only values above a threshold lead to a network connection.

For establishing structural connectivity, diffusion tensor imaging (DTI) or diffusion spectrum imaging (DSI) can be applied. Using deterministic tracking, for example, the number of streamlines between brain regions can be represented in a matrix. For probabilistic tracking, matrix elements would represent the probability to reach a target node starting from a source node. In both cases, the weighted matrix can either be analyzed directly or be thresholded so that connections are only formed if a minimum number of streamlines or a minimum probability has been reached.

The choice of nodes and edges can be influenced by the anatomical parcellation schemes and measures for determining connectivity (Rubinov and Sporns, 2010; Eickhoff, Yeo, and Genon, 2018). This choice must be carefully considered as different choices might not only change the topology by removing or adding a few nodes or connections but might alter the local and global network features that will be discussed in the following sections.

Connections of a network can be binarized or weighted. Binary connections only report the absence or presence of a connection. Weighted links can also show the strength of a connection. For structural connectivity, weights can indicate the number of fibers between brain regions (e.g., the streamline count of deterministic tracking), the degree of myelination, the probability that a node can be reached from another node (e.g., probabilistic tracking), or the amount of dye traveling from one node to another (traditional tract-tracing studies). For functional connectivity, weights can indicate the correlation in the time course of signals of different nodes.

2.2 Topological Features

Local Scale—Single Node Features

Networks can be characterized at different levels, ranging from properties characterizing a whole network at the global scale to properties of network components at the local scale. Starting from the local scale, components of a network are its nodes and

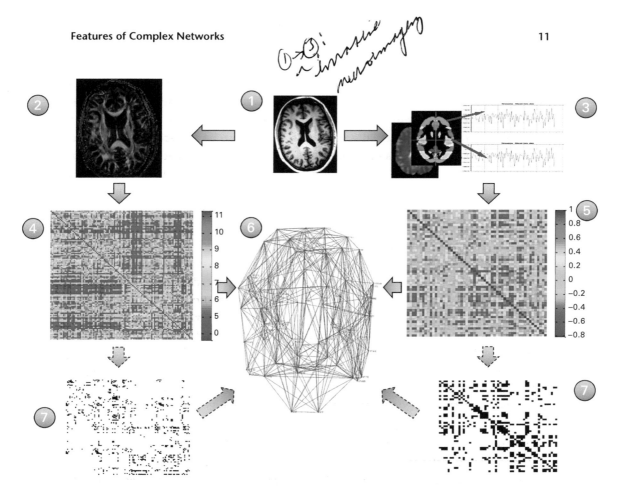

Figure 2.2
Workflow for structural and functional connectivity analysis. High-resolution anatomical magnetic resonance imaging scans of each subject are used as references for further measurements (1). For establishing functional connectivity, a time series of brain activity in different voxels or regions can be derived (3). The correlation between the time series of different voxels or, using aggregated measures, brain regions can be detected and represented as a correlation matrix (5). This matrix can either directly be interpreted as a weighted network (6) or can be binarized in that only values above a threshold lead to a network connection (7). For establishing structural connectivity, diffusion tensor imaging or diffusion spectrum imaging can be applied (2). Using deterministic tracking, for example, the number of streamlines between brain regions can be represented in a matrix (4). This weighted matrix can either be analyzed directly (6) or be thresholded so that connections are only formed if a minimum number of streamlines has been reached (7).

edges. Edges can be weighted, taking continuous (metric) or discrete (ordinal) values indicating the strength of a connection. Alternatively, they could just have binary values with zero for absent and one for existing connections.

When one thinks about neural systems, there could also be multiple edges between two nodes—for example, a fiber bundle connecting two brain regions. However, such multigraph networks are usually simplified in that the number of fibers is either neglected (binary values) or included in the strength of a connection. In addition to fiber count, one might also think of other properties of connections such as delays for signal propagation or degree of myelination. Whereas such properties likely have significant impact on network function, they are currently not part of the analysis of network topology but are increasingly part of models for network dynamics.

The other component at the local scale is a network node. A node could be a single neuron but, as for edges, could also be an aggregate unit of neurons such as a population or a brain area. The *degree* of a node is the sum of its incoming (afferent) and outgoing (efferent) connections. The number of afferent and efferent connections is also called the *in-degree* and *out-degree*, respectively. When k_i denotes the degree of the node i of a network with N nodes, the series $(k_1, \ldots, k_N)$ with increasing degrees is called the *degree sequence* of the network. Nodes with a high number of connections, that is, a large degree, are called *network hubs*. For structural and effective connectivity, the ratio between the in- and out-degree of a node can give information about its function: nodes with predominantly incoming connections can be seen as integrators (convergence) whereas nodes with mainly outgoing connections can be seen as distributors (divergence) or broadcasters of information. These distinctions can be useful when nodes are otherwise similar, for example, distinguishing different types of network hubs.

For undirected networks, every connection between nodes is bidirectional (e.g., for functional networks measuring correlation). For such networks, if a node A is connected to a node B with a bidirectional link, this link is counted as one connection when calculating the degree of the node. Likewise, it does not make sense to distinguish in-degree and out-degree as they will have the same values.

Local measures can also refer to the neighborhood of a node. All nodes that directly project to a node or directly receive projections from that node are called *neighbors* of that node. The connectivity between neighbors is used to assess local clustering. The ratio of the number of existing edges between neighbors and the number of potential connections between neighbors forms the *local clustering coefficient* (see figure 2.3A), a measure of neighborhood connectivity.

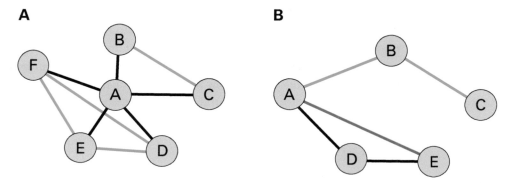

Figure 2.3
Local and global measures. (A) Local clustering coefficient: All nodes that are connected with node A are neighbors of that node. The local clustering coefficient of node A is the number of connections between neighbors (green edges) divided by the number of all potential connections between its neighbors. In this case, the local clustering coefficient is $C_A = 4/10 = 0.4$, meaning that 40% of connections between neighbors exist. (B) Shortest paths: The shortest path is the path between two nodes with the lowest possible number of connections in the path. The *path length* of that path is the number of connections that have to be crossed to go from one node to another. In this example, the length of the shortest path between nodes A and C (A→B→C) is 2 and the length of the shortest path between nodes A and E (A→E) is 1.

The local clustering coefficient for an individual node i with k_i neighbors (node degree k_i) and Γ_i directed edges between its neighbors is

$$C_i = \frac{\Gamma_i}{k_i(k_i - 1)}.$$

Note that this equation is not defined (division by zero) if a node is isolated (no neighbors) or has only one neighbor (consequences and potential solutions to this problem are discussed in Kaiser, 2008).

Another local measure of a single node is how many shortest paths (see figure 2.3B) are containing that node. The shortest path between two nodes is the length of the path with the lowest possible number of connections. For counting how many paths pass through a node, the shortest paths between all pairs of nodes are calculated. It is then counted how many shortest paths include a certain node. Note that this measure is listed as a local measure as it is an attribute of a single node even though information of the whole network is used to calculate this measure. For comparing this measure for two nodes of the same network, it does not matter whether the absolute number of shortest paths containing one node (stress centrality) or the relative frequency of how often that node is part of shortest paths (betweenness centrality) is taken. This measure of how frequently a node is part of shortest paths is called *node betweenness*. In a similar way, the frequency of shortest paths containing a certain edge is the *edge betweenness* of that edge. There are numerous kinds of centrality (chapter 3 in Brandes and Erlebach, 2005) in addition to measures of how many shortest paths run through network components—for example, closeness centrality, which is the reciprocal of the total distance of a node to any other node of the network, and degree centrality, which is the degree of a node.

Global Scale—Aggregate Measures

We are now zooming out of a network and observe properties that characterize the network as a whole, leaving out the intermediate regional scale for a moment. Whereas local measures look at properties of individual components, global measures at the macroscale look at the whole network. This is useful when comparing a given neural network with artificially generated networks called *benchmark networks*. It is also useful when comparing neural networks from different species or the same species at different levels of organization (area, column, layer). In these cases, the number of nodes and edges as well as their identity (e.g., comparing networks that contain a certain region with networks that do not) might differ; however, aggregate measures can still be used to detect changes at the macroscale.

As mentioned earlier, the edge density, sometimes called connectivity, of a network is the proportion of connections that exists relative to the number of potential connections of a network. For a directed network with N nodes, each node can connect to at most $N-1$ other nodes. Therefore, the edge density of a network with E edges and N nodes is

$$d = \frac{E}{N(N-1)}.$$

For an undirected network, the edge density becomes

$$d = \frac{2E}{N(N-1)};$$

note the factor 2 in the numerator so that any potential undirected edge between two nodes is only counted once and not twice. An edge density of 1, corresponding to a percentage of 100%, would mean that all potential edges exist. In biological networks, however, only a small fraction of potential connections occurs. For the corticocortical fiber tract connectivity of the mammalian brain, for example, the edge density ranges between 10% and 30%. For the connectivity between neurons in the nematode *C. elegans*, the edge density is 3.85%.

The edge density gives a first indication of how well-connected a network is. However, the number of steps it takes to go from one node to another might still vary considerably depending on the topology of the network. A measure of traveling through a network is the number of connections one has to cross, on average, to go from one node to another. Formally, this *average shortest path* (also called all-pairs shortest path) of a network with N nodes is the average number of edges that have to be crossed on the shortest path from any one node to another:

$$ASP = \frac{1}{N(N-1)} \sum_{i,j} d(i,j) \text{ with } i \neq j$$

where $d(i,j)$ is the length of the shortest path between nodes i and j having as few intermediate steps between nodes i and j as possible. Note that the definition for the *characteristic path length L* is slightly different (Watts, 1999) in that for each node the average shortest path length to any other node is calculated and the median, instead of the mean, value over all nodes is returned as L.

In practice, in particular for networks with directed edges, there might be several pairs of nodes for which no path exists. In such cases, graph theory would demand setting the distance $d(i,j)$ between the two nodes to infinity. However, having only one such pair in a network would give an L of infinity as well. In practice, such infinite values are excluded; that means the average shortest path only takes the average of existing shortest paths between pairs of nodes. Alternatively, a measure called global efficiency can be used (Latora and Marchiori, 2001; Achard and Bullmore, 2007). Global efficiency uses a sum of the inverse of the distance L so that nonexisting paths, leading to infinite distance, contribute a zero value to the sum:

$$E_{global} = \frac{1}{N(N-1)} \sum_{i \neq j} \frac{1}{d(i,j)}$$

where $d(i,j)$ is the length of the shortest path between nodes i and j and N is the number of nodes.

Another aggregate measure based on the local features of individual nodes is average neighborhood connectivity—the (global) clustering coefficient. The clustering coefficient is just the average of the local clustering coefficient C_i of all nodes:

$$C = \frac{1}{N} \sum_i C_i$$

In relation to the characteristic path length as global efficiency (how well are any two nodes of a network connected?), the clustering coefficient can also be called *local efficiency* (how well are neighbors of a node connected?) (Latora and Marchiori, 2001; Achard and Bullmore, 2007).

The clustering coefficient will increase whenever the edge density increases as a higher probability that any two nodes are connected also means that connections between neighbors are more likely. Therefore, either comparing clustering coefficients for networks with different edge densities should be avoided, or they should use a normalized coefficient. Such a normalized coefficient would be the clustering coefficient relative to the clustering coefficient of a random network. However, normalization does not work if one of the two compared networks has a much higher edge density: whereas the clustering coefficient of a network with low edge density can be 2 to 3 times as high as the edge density of that network, the clustering coefficient of a network with, say, 60% edge density can never be 2 to 3 times as high as the edge density.

Regional Scale—Groups of Network Nodes

Often, we are interested in an intermediate level of organization going beyond single nodes but not including the whole network. At this scale, we can observe measures for subsets of nodes which share similar connections—for example, dealing with visual input. Such measures look at the sets of nodes where the connectivity within the set is larger than between the set of nodes and the rest of the network. Such sets of nodes are called clusters, modules, or, following social network analysis, communities.

Clusters

Clusters or modules are parts of a network with many connections (high edge density) within such a part and few connections (low edge density) to the remaining nodes of the network. There are many different algorithms to detect clusters of a network (Hilgetag, O'Neill, and Young, 2000; Girvan and Newman, 2002; Palla et al., 2005). As a general rule, algorithms can be distinguished on the basis of three features (many other classifications exist). First, algorithms could lead to hierarchical or nonhierarchical solutions. Hierarchical solutions identify not only modules but also submodules within modules, sub-submodules within submodules, and so on (Clauset et al., 2008). That means that the modular organization at different hierarchical levels can be observed.

Note that the hierarchical parcellation of the network critically depends on the threshold that one chooses. Whereas an early threshold, near the root of the tree, might only show the main clusters—just one hierarchical level—a late threshold, close to the leaves of the tree, might result in numerous subclusters that are so small that the distinction between clusters becomes weak.

Second, algorithms can use a predefined number of clusters which need to be detected or can determine the number of clusters themselves. If the number of clusters is known, algorithms similar to *k*-means, where *k* is the number of clusters, can be used to detect the clusters. However, in many biological applications, the number of clusters is not known beforehand, and algorithms that determine the number during the clustering process are needed.

Third, algorithms can lead to overlapping or nonoverlapping cluster-classifications of nodes. For nonoverlapping algorithms, a node will belong to one and only one cluster of the network. However, this assignment to a cluster can often be ambiguous with only a slightly lower preference for assigning the node to another cluster. For overlapping algorithms, a node can belong to several clusters with different likelihoods. Say a node could belong to clusters A, B, and C with likelihoods of 20%, 30%, and 50%, respectively. Such overlapping cluster memberships for cortical nodes can point to nodes that integrate information from several modules—for example, from the visual and auditory system.

An example of a modular organization is human corticocortical connectivity (see figure 2.4A), based on DSI (Hagmann et al., 2008). Note that nodes in the same cluster, having a high structural similarity, also have a similar function. The cluster architecture of the same network can also be represented by a dendrogram using hierarchical clustering (see figure 2.4B).

A measure that has received a lot of attention for topological clusters in recent times is modularity. Modularity (Q) is a reflection of the natural segregation within a network (Newman, 2004) and can be a valuable tool in identifying the functional blocks within. The complex networks modularity measure Q can be used to assess how well a parcellation into nonoverlapping modules represents the modular architecture of a network. Given two parcellations into distinct modules for the same network, the parcellation with the higher value of Q would be preferred. So how can the modularity Q be computed? Given a parcellation that assigns to each node i a label c_i identifying to which module the node belongs, the modularity Q is the difference between the number of edges that lie within a community in the actual network and a random network of the same degree sequence. A high level of topological clustering is reflected in a high value of modularity. The modularity for a directed network is given by the following (Newman, 2006):

$$Q = \frac{1}{m} \sum_{i,j} \left[a_{ij} - \frac{k_i^{in} k_j^{out}}{m} \right] \delta_{c_i, c_j}$$

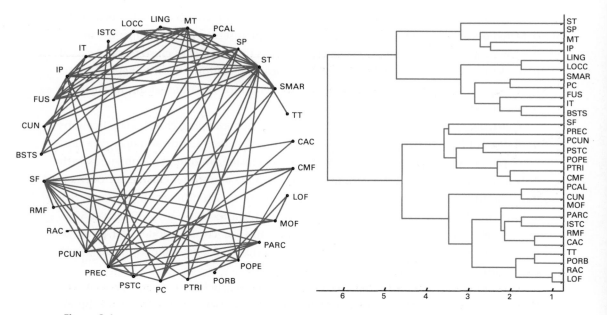

Figure 2.4
Clusters. Left: Cluster structure of human corticocortical connectivity, based on diffusion spectrum imaging (Hagmann et al., 2008). Cortical areas were arranged around a circle by evolutionary optimization, so that highly interlinked areas were placed close to each other. Note that nodes in the same cluster, having a high structural similarity, also have a similar function. Right: Dendrogram of the same network using hierarchical clustering (for acronym definitions of brain regions see figure 1 of Hagmann et al., 2008). A dendrogram running from the root to the leaves (here: from left to right) consists of branches connecting objects in the tree. The distance of the branching point on the x-axis is the rescaled distance when clusters are combined.

where m: total number of edges in the network (note that bidirectional links are counted twice); a_{ij}: element of adjacency matrix; k_i^{in}: in-degree of node i; k_j^{out}: out-degree of node j; δ_{c_i, c_j}: Kronecker delta (only one if nodes i and j are in the same module and zero otherwise) with c_n being the label of the module to which node n belongs.

This measure can be used as a cost function in cluster algorithms where the aim is to maximize the modularity function Q. Clusters can be defined not only by grouping nodes but also by grouping edges into link communities (Ahn et al., 2010).

A measure to quantify to what extent a node communicates with nodes in other modules is the participation coefficient (Guimera and Amaral, 2005). The participation coefficient P_i measures how "well distributed" the links of node i are among different modules:

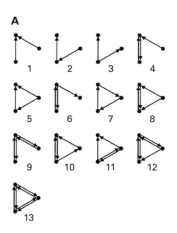

Brain Network	ID	Real	Random
Human Cortex	13	N/A	N/A
Macaque Visual Cortex	9	410	121.55 (21.03) z = 13.79
Macaque Cortex	9	1833	223.66 (34.99) z = 46.22
Cat Cortex	9	1217	472.33 (52.85) z = 14.16
C. elegans	4	2999	1067.03 (121.52) z = 15.98
	6	3415	1164.31 (134.71) z = 16.79

Figure 2.5
Network motifs. (A) Overview of all 13 possible ways to connect three nodes (patterns with isolated nodes are not considered). (B) Three-node patterns that occur significantly more often in the human (Iturria-Medina et al., 2008), macaque, cat (connectivity between regions) as well as *C. elegans* (connectivity between neurons) structural connectivity than in rewired networks and are thus network motifs (adapted from Milo et al., 2002; IDs refer to the numbers in part A).

$$P_i = 1 - \sum_{s=1}^{N_M} \left(\frac{K_{is}}{k_i} \right)^2$$

where K_{is} is the number of links of node i to nodes in module s, where N_M is the number of modules, and k_i is the total degree of node i. The participation coefficient P_i is close to one if its links are uniformly distributed among all the modules, and zero if all its links are within its own module.

Network motifs

Modules are relatively large structures comprising tens or hundreds of nodes. Modules are often linked to function in that nodes of the same module tend to have a similar function. However, there could also be smaller subgraphs with only a few nodes that could have a specific function for a network. For a subgraph with only two nodes A and B, there are three ways in which directed edges could exist between them: a connection in one direction, A→B; a connection in the reverse direction, A←B; and a bidirectional connection, A↔B. For subgraph counting, the case that no connections between the nodes exist is not taken into account. Also, the identity of the nodes is not retained; therefore, A→B and A←B are treated as one pattern. For three nodes, there are already 13 different patterns of how directed edges could be distributed (see figure 2.5A). For a real-world network it is then possible to count how often each potential two-node or three-node pattern occurs. If a pattern occurs significantly more often than in a randomly organized network with the same degree distribution, it is called a network *motif* (Milo et al., 2002).

So when does a pattern occur significantly more often than would be expected for a random organization? To decide this, a set of benchmark networks is generated where the number of nodes and edges is identical but, starting from the original network, edges are rewired while each node maintains its original in-degree and out-degree. Thus, the degree distribution of the network remains unchanged. This means that each node still has the same degree after the rewiring procedure but that additional information—for example, the cluster architecture—is lost. In the next step, for each benchmark network the number of occurrences of a pattern is determined. Then, the pattern count of the real-world network can be compared with the average pattern count of the benchmark networks; patterns that occur significantly more often in the real-world network than in the benchmark networks are called network motifs. Figure 2.5B shows characteristic network motifs of size three for different neural networks (Milo et al., 2002).

Finding these motifs is a computationally hard problem: as the size of the motifs gets bigger, the time needed to calculate grows exponentially (for a survey of motif detection strategies, see Ribeiro et al., 2009) even though fast algorithms exist (Aparício et al., 2016). There is also a conceptual problem. The benchmark for counting the number of motifs is a rewired network with the same degree distribution. Even if the degree distribution remains identical, random rewiring removes the topological cluster architecture of the real network. Using a rewiring that maintains both the degree distribution and the topological cluster architecture leads to a lower number of network motifs: many motifs, such as highly connected three- or four-node motifs, are frequent when densely connected modules exist. If the rewired brain network contains such densely connected modules as well, those patterns do not occur significantly more often than in the original network. Therefore, only few motifs for the networks considered in Sporns and Kötter (2004) remain.

Types of Networks

We already mentioned that network measures can be used to compare networks with each other. Often, we are interested not only in how network measures differ but in whether the *type* of network differs. Although each neural network has a unique topological and spatial organization, such types or classes of networks can be used for classification and comparison (see figure 2.6). Such classes are based on global features of

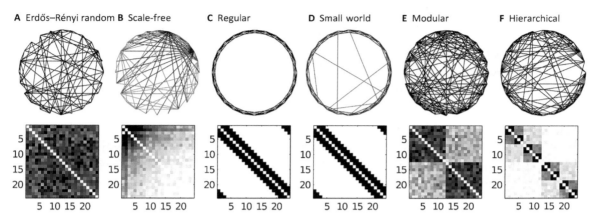

Figure 2.6

Types of networks. Networks contain 24 nodes and 142 undirected edges. The top panel shows individual networks where nodes are located on a circle. The bottom panel shows the average probability over 100 networks to display a connection in the adjacency matrix: white denotes edges that are always absent whereas black shows edges that are always present. (A) Erdős–Rényi random network. (B) Scale-free network with dark edges (top panel) indicating highly connected nodes or hubs. (C) Regular or lattice network with high connectivity between neighbors. (D) Small-world network with blue edges (top panel) representing shortcuts of the network. Note that for the average probability plot (bottom panel), shortcuts are invisible due the averaging over 100 networks. (E) Modular network with two modules. (F) Hierarchical network with two modules consisting of two submodules each. Thus, there are two hierarchical levels of organization.

the degree distribution and the community organization. The following subsections show different types and their characteristic properties. Note that real-world networks, however, might show a combination of different classes—for example, being modular and small world.

Random networks

Whereas many networks are generated by a random process, the term random network normally refers to the type of Erdős–Rényi random networks (Erdős and Rényi, 1960). Random networks are generated by establishing each potential connection between nodes with a probability p. This probability, for a sufficiently large network, is then equivalent to the edge density of the network—that is, the connection density. The process of establishing connections resembles flipping a coin where an edge is established with probability p and not established with probability $q = 1 - p$. Therefore, the distribution of node degrees follows a binomial probability distribution. For large numbers of nodes and a low probability p, the probability $P(k)$ that a node has k connections can be approximated by a Poisson distribution, and hence the term "exponential degree distribution" is also used (Bollobas, 1985). The distribution can be shown as a histogram where the counts for the different bins are plotted as data points. For the "exponential" degree distribution, $P(k) \sim e^{-k}$, of a random network, points are arranged on a line for a logarithmic plot of $\log(P(k))$.

Scale-free networks
Scale-free networks are characterized by their specific distribution of node degrees. The degree distribution follows a power law where the probability that a node with degree k exists, or the frequency of occurrences for real-world networks, is given by $P(k) \sim k^{-\lambda}$. This is different from the random networks discussed above where the degree distribution follows an exponential distribution, $P(k) \sim e^{-k}$. The exponent of a power-law degree distribution can vary depending on the network which is studied. For example, for functional connectivity between voxels in human MRI, λ was found to be 2.0 (Eguiluz et al., 2005) whereas λ was 1.3 for functional connectivity of neurons in the hippocampus determined through calcium imaging (Bonifazi et al., 2009).

Data points for a power-law degree distribution lie on a straight line for a log-log plot of $\log(P(k))$ against $\log(k)$. To test this power-law relationship, the cumulative distribution $P_c(k) = P(X > k) = 1 - F(k)$ with $F(k) = P(X \leq k)$ where X is the number of connections of a node is plotted (see figure 2.7C). As the histogram uses the same bin widths, the bins for high degrees have fewer entries than the bins for nodes with low degree values. Therefore, data points in the histogram will fluctuate more strongly for the tail of the distribution. In addition to the visual inspection of the log-log plot, a statistical analysis is needed to test for a power-law behavior (Clauset et al., 2009). Such an analysis will determine the goodness-of-fit with a power-law distribution and will also compare this result with alternative distributions.

Previous studies have shown that functional networks of the human brain, looking at signal correlations between voxels in functional magnetic resonance imaging (fMRI), are scale-free (Eguiluz et al., 2005). However, at the gross level of signal correlations between brain regions, it was argued that these functional networks are not scale-free (Achard et al., 2006).

There are two potential problems with determining whether a network shows a power-law degree distribution and is thus scale-free. One problem arises when only part of the network is known. For example, one might like to test scale-free properties of neuronal networks but has connectivity only within one column, not with other parts of the brain. In this case, connectivity is known for only a subset or sample of the nodes of the whole network. Using such incomplete sampling, is it still possible

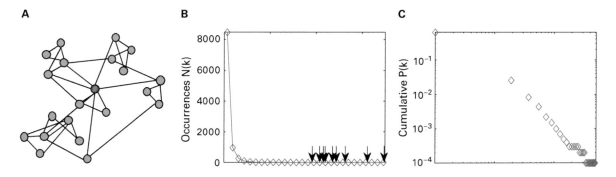

Figure 2.7
Scale-free networks. (A) Scale-free networks contain highly connected nodes or hubs (dark node in this example). (B) Degree distribution of a scale-free network with 10,000 nodes and 20,145 connections. In contrast to random networks with one characteristic scale where all node degrees k are close to the average, scale-free networks can contain nodes with degrees that are several standard deviations away from the average. In this example, there are 13 nodes with degrees that are nine standard deviations away from the average degree of four (arrows); the maximum degree is 504 (beyond the figure axes). (C) The cumulative frequency $P(k)$ that a node with degree k occurs in the network follows a power law leading to a straight line in a bilogarithmic plot.

to test whether the whole network is scale-free or not? Unfortunately, the amount of unknown or not included connections or nodes might change the shape of the degree distribution (Stumpf et al., 2005); in particular, missing nodes could mean that the rarely occurring hubs are not part of the sample, resulting in classifying a scale-free network as an Erdős–Rényi random network.

Another problem is networks with a low number of nodes, where the degree distribution consists of only one or two orders of magnitude. Unfortunately, this is the case for regional brain connectivity where networks consist of 100 or fewer nodes: structural networks in the macaque, cat, and human within one hemisphere usually consist of around 30–100 nodes. Such a low number of nodes results in power-law fits that are not robust. For such networks, individual outlier nodes with very high connectivity might directly alter the tail of the degree distribution, whereas for larger networks the degrees of several nodes will be binned so that outliers are less influential. However, we can test for scale-free behavior by using indirect measures whose outcome is not altered significantly by one individual node (Kaiser, Martin, et al., 2007): when comparing the effect of node removal in the cat and macaque structural connectome, targeting hubs or removing nodes randomly, the change in path length observed in brain networks was most similar to the pattern observed in scale-free benchmark networks. So even though the degree distribution cannot be tested, the robustness after simulated lesions is most similar to that of a scale-free network.

Small-world networks

Many networks exhibit properties of small-world networks (Watts and Strogatz, 1998). The term small world refers to experiments in social networks by Stanley Milgram where a person could reach any other person through a relatively short chain of acquaintances, the "six degrees of separation" (Milgram, 1967). However, relatively short does not mean that the average number of connections to cross from one node to another, the characteristic path length, is minimal. Indeed, the path length is usually higher than for Erdős–Rényi random networks with the same number of nodes and edges. However, connectivity between node neighbors, the clustering coefficient, is much higher than for random networks.

So when can a network be considered a small-world network? Unfortunately, there is no clear criterion. In general, to classify a network as small world, its clustering coefficient should be much higher than the clustering coefficient of Erdős–Rényi random networks. For Erdős–Rényi random networks, the clustering coefficient has the same value as the edge density (connections between neighbors are as likely as any other connections of the network), so edge density might be used for the comparison. In addition, the characteristic path length of the network should be comparable to that of a random network that means slightly but not excessively higher than that value.

A measure to summarize to what extent a network shows features of a small-world network is small worldness $S = (C / C_{\text{rand}}) / (L / L_{\text{rand}})$ where C is the clustering coefficient and L is the characteristic path length of an observed network and a random network (Humphries and Gurney, 2008). Note that this measure is useful for comparing small-world networks but not sufficient for determining whether a network is a small-world network or not: a high value of S might also occur for networks with extremely high characteristic path length as long as the clustering coefficient is much higher than for random networks.

Note that a high clustering coefficient does not necessarily mean that a network contains multiple clusters! Indeed, the standard model for generating small-world networks by rewiring regular networks (Watts and Strogatz, 1998) does not lead to multiple clusters. In addition, small-world and scale-free properties are compatible, but not equivalent; a network might be small world but not scale-free and vice versa.

Small-world properties were found on different organizational levels of neural networks: from the tiny nematode *C. elegans* with about 300 neurons (Watts and Strogatz, 1998) over cortical structural connectivity of the cat and the macaque (Hilgetag, Burns, et al., 2000; Sporns et al., 2000; Hilgetag and Kaiser, 2004) to human structural (Hagmann et al., 2008) and functional (Achard et al., 2006) connectivity.

Modular and hierarchical networks

Two central topological features of brain networks, in particular of the cerebral cortex, are their modular and hierarchical organization. Modular networks consist of multiple clusters. If these clusters occur at different levels, a cluster consisting of multiple subclusters, subclusters consisting of several sub-subclusters, and so on, the network can be called a hierarchical modular network. Note that for only one level, the network would be modular but not hierarchical. On the other hand, networks that are hierarchical but not modular seem impossible.

A modular hierarchical organization of cortical architecture and connections is apparent across many scales, from cellular microcircuits in cortical columns (Mountcastle, 1997; Binzegger et al., 2004) at the lowest level, via cortical areas at the intermediate scale, to clusters of highly connected brain regions at the global systems level (Hilgetag, Burns, et al., 2000; Breakspear and Stam, 2005; Kaiser, Görner, and Hilgetag, 2007). The precise organization of these features at each level is still unknown, and there exists controversy about the exact organization or existence of modules even at the level of cortical columns (Horton and Adams, 2005; Rakic, 2008). Nonetheless, current data and concepts suggest that at each level of neural organization clusters arise, with denser connectivity within than between the modules. This means that neurons within a column, area, or cluster of areas are more frequently linked with each other than with neurons in the rest of the network.

Several potential biological mechanisms for generating hierarchical modular networks have been described (see chapter 9). An algorithm for generating hierarchical network with multiple level and multiple modules at each level can be found at https://mitpress.mit.edu/changing-connectomes.

2.3 Spatial Features

The previous section looked at topological properties of neural networks, but brain networks also have spatial properties in that each node and edge has a three-dimensional location and extension, be it volume for nodes or diameter and trajectory for edges. Given the spatial extent of network components, space is often a limiting factor for the structural organization of neural systems. For example, all-to-all connectivity between all neurons of the brain is impossible given the limited volume available for white matter fiber tracts within the skull.

Neural networks that extend in two or three dimensions can be described as spatial networks (Watts, 1999). However, the concept of space could be extended to nonmetric space. For example, using an ordinal scale, entities in the same region could be assigned a distance of 0, in adjacent regions a distance of 1, and—with one neighbor in between—a distance of 2, and so forth. One example for this is the yeast protein-protein interaction network where there are relatively more interactions within the mitochondrium ("distance 0") than between mitochondrium and other cellular compartments (Schwikowski et al., 2000).

In theoretical studies, spatial graphs are usually generated by distributing the nodes randomly in space. After the nodes are arranged, edges between the nodes are added to the network either randomly or depending on the distance between the nodes. Additionally, connected nodes could be drawn together by an *a posteriori* pulling algorithm, resulting in spatial clusters of connected nodes (Segev et al., 2003). Note that these approaches for generating spatial networks miss important features of real-world networks such as a small-world, hub, or modular organization, and I will discuss biologically more realistic models in part II.

Connection Length Distributions

Each node and edge in neural networks, at least after potential migration during development, has a constant spatial position. Such a spatial layout is far from random, but to what extent self-organization or genetic predisposition determines location is still an open question. One first step in observing the spatial organization of a neural network is to look at the lengths of connections. If two nodes are connected, the Euclidean distance between the positions of both nodes can be a lower bound of the length of the connection. Note that even for cortical fiber tracts this gives a reasonable approximation: for the prefrontal cortex in the macaque only 15% of the connections are strongly curved, and dense fibers, in particular, tend to be completely straight (Hilgetag and Barbas, 2006). It is often interesting to observe how many connections go to nearby targets and how many extend over a long distance, potentially linking different components of the neural network. This can be readily observed using a histogram of the connection lengths of a network. These histograms for anatomical connection lengths, ranging from *C. elegans* and rat neuronal to macaque and human cortical connectivity, all show a decay of the frequency over distance: short-distance connections are more frequent than long-distance connections (see figure 2.8). For these systems, the distribution can best be approximated through a Gamma distribution (Kaiser et al., 2009).

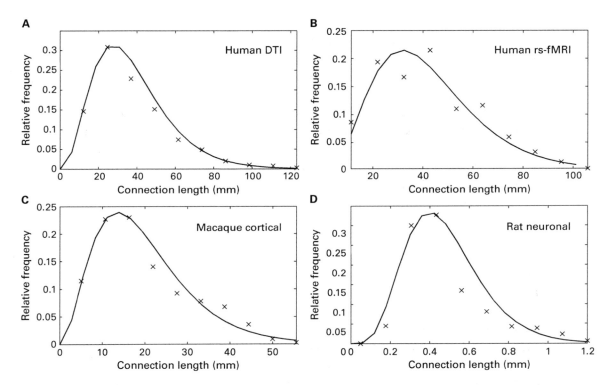

Figure 2.8
Connection lengths in regional and neuronal networks. Connection length distributions where the relative counts of a histogram are plotted as data points (x) fitted with a Gamma function (solid line). For cortical networks, only connectivity within a hemisphere was considered. Despite different species, levels of organization, and types of connectivity, all distributions show an early peak and a later distance-dependent decay in the frequency of connection. (A) Human diffusion tensor imaging (DTI) network between 55 brain regions where each existing fiber tract is represented by one (unweighted) network connection. (B) Human resting-state functional magnetic resonance imaging (rs-fMRI) network between 55 brain regions where the top 20% of correlation are represented by a network connection (that means the threshold was set so that the edge density was 20%). (C) Macaque cortical fiber tract network of 95 brain regions. (D) Rat supragranular pyramidal cell neuronal network of layers II and III of the extrastriate visual cortex. (Plots C and D adapted from Kaiser et al., 2009.)

Component Placement Optimization

Establishing connections involves metabolic structural costs for building connections (especially for myelinated axons) as well as dynamic costs for transmitting action potentials and reestablishing the resting membrane potential through active transport of ions. It is therefore natural to assume that these energy costs should be as low as possible (Cherniak, 1992; Chklovskii et al., 2002; Wen and Chklovskii, 2008). One possibility for reducing costs is to have a lower number of long-distance connections. The frequency of such long-distance connections, relative to short-distance connections, can be seen in the connection length histogram. One can also test how far away the combined length of all connections together, the total wiring length, is from the shortest possible total wiring length. Such an optimal solution can be found in two ways: rearranging the connections of each node whereas the position of a node remains the same or rearranging the position of each node, swapping around the position of nodes, while retaining the connectivity of each node projecting to the same target nodes (not target positions).

Reducing the total length by reordering connections could lead to minimal wiring of a system. One possibility would be to establish connections ranked by their length, connecting nodes that are closest to each other first. However, this could result in a fragmented network where parts of the network are unreachable from many starting nodes. To secure reachability, one could start with a minimum spanning tree (see the glossary in this volume for explanations of terms) that connects N nodes with $N - 1$ edges and a minimal wiring length (Cormen et al., 2009), and then add remaining connections again using short-distance connections as described before.

Alternatively, wiring length reductions in neural systems can be achieved by suitable spatial arrangement of the components. Under these circumstances, the connectivity patterns of neurons or regions remain unchanged, maintaining their structural and functional connectivity, but the layout of components is perfected such that it leads to the most economical wiring. In the sense of this "component placement optimization" (Cherniak, 1994), any rearrangement of the position of neural components, while keeping their connections unchanged, would lead to an increase of total wiring length in the network. For a small number of nodes, all possible arrangements of their positions can be tested. For larger networks, however, such an approach is not feasible: the number of possible layouts for N nodes is $N!$ (e.g., 10^{148} possibilities for 95 nodes). In those cases, optimal solutions can be approximated only through numerical routines such as simulated annealing (Metropolis et al., 1953) or others. Studies on neuronal networks in *C. elegans* and on cortical networks in the macaque have shown that a reduction by 50% and 30%, respectively, is possible (Kaiser and Hilgetag, 2006).

The Role of Long-Distance Connections

Neural networks contain a substantial proportion of long-distance projections, many more than minimally rewired networks of the same size. Due to these far-reaching spatial shortcuts, minimal rewiring of the biological networks led to a global reduction in the amount of wiring, as well as to a reduced metric length of the average shortest path between components. Moreover, minimally rewired networks showed increased clustering of connections within local neighborhoods (Kaiser and Hilgetag, 2006). Such features are potentially beneficial for the organization of neural networks, resulting in economical wiring as well as greater local integration of network nodes.

However, minimal network rewiring also resulted in significantly increased average path length (corresponding to the number of connection steps in the shortest pathways) between the components. Thus, it appears that in biological neural networks economical wiring and tight integration of local components are counterbalanced by the global minimization of processing steps across the network (cf. Karbowski, 2001). Indeed, when evaluated on a comparative scale for minimal wiring and minimal path lengths, the neural networks considered here were placed further away from the optimal configuration for minimal wiring than from the best network arrangement for minimizing processing steps (see figure 2.9).

The importance of network shortcuts for reducing the number of processing steps has been pointed out before (e.g., Striedter, 2005), particularly in the context of small-world network architectures (Watts and Strogatz, 1998). In brain networks, these network shortcuts are mainly formed by long-distance connections. This conclusion, while intuitive, is not trivial, as one could also imagine alternative scenarios in which network shortcuts arise from short-distance connections—for example, connections between spatially nearby regions that belong to different topological modules (Kaiser and Hilgetag, 2006). The coincidence of long-distance connections with network shortcuts hints at a close match between the spatial layout and topology of neural networks (Young and Scannell, 1996). It will be an interesting task for future studies to explore more fully the developmental and evolutionary reasons for this coincidence.

Minimizing average path length, that is, reducing the number of intermediate transmission steps in neural integration pathways, has several functional advantages. First, the number of intermediate nodes which may introduce interfering signals and noise is limited. Second, by reducing transmission delays from intermediate connections, the speed of signal processing and, ultimately, behavioral decisions is increased. Third, long-distance connections enable neighboring as well as distant regions to receive activation nearly simultaneously (Kaiser and Hilgetag, 2004a; Masuda and Aihara, 2004) and thus facilitate synchronous information processing in the system (cf. von der

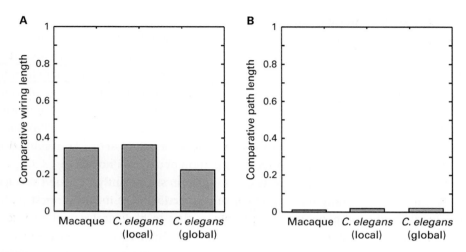

Figure 2.9
Wiring arrangement of neural networks compared to minimum and maximum case benchmark networks. (A) Actual total wiring length relative to the minimum wiring length solution (value zero, yielded by simulated annealing of component positions) and to networks optimized for maximum total wiring length (value one, also yielded by simulated annealing). The wiring of the different neural networks lies close to the middle between minimum and maximum case component arrangements. (B) Average shortest path length (characteristic path length) in neural networks relative to networks optimized for minimum path length (value zero, yielded by simulated annealing of wiring organization) and maximum path length (value one). Actual path lengths in the neural networks are close to the lower bound of networks optimized for minimum paths (adapted from Kaiser and Hilgetag, 2006).

Malsburg, 1995; König et al., 1995). Fourth, the structural and functional robustness of neural systems increases when processing pathways (chains of nodes) are shorter. Each further node introduces an additional probability that the signal is not transmitted, which may be substantial (e.g., failure rates for transmitter release in individual synapses are between 50% and 90%; Laughlin and Sejnowski, 2003). Even when the signal survives, longer chains of transmission may lead to an increased loss of information. A similar conclusion, on computational grounds, was first drawn by John von Neumann (1958) when he compared the organization of computers and brains. He argued that, due to the low precision of individual processing steps in the brain, the number of steps leading to the result of a calculation ("logical depth") should be reduced and highly parallel computing would be necessary. As we will see later, a loss of long-distance connections might also underlie pathological changes in neural network function.

The advantages resulting from a short average number of processing steps appear as least as beneficial for the organization and function of neural networks as those commonly cited for wiring minimization. Previous analyses of cortical organization have demonstrated that the convoluted, laminar architecture of the mammalian cerebral cortex and the segregation into gray and white matter reduce total white matter volume and shorten projection lengths (Murre and Sturdy, 1995), also reducing conduction delays (Wen and Chklovskii, 2005). Ultimately, these adaptations point in the same direction as reductions in the number of processing steps, toward maximizing information processing speed. Therefore, it is plausible that neural systems are adapted to more than just one design constraint, and that their observed organization is the outcome of an optimization of multiple parameters, which may be partly opposed to each other. For cortical networks, for example, additional constraints may arise from spatial factors that limit growth (Kaiser and Hilgetag, 2004a, 2004b) or critical periods for the establishment of cortical areas and their interconnections (Rakic, 2002). In addition to current ontogenetic constraints, the evolutionary history of a neural network might also conserve features of its predecessors, some of which may not be optimal for the present system (Striedter, 2005).

Lateralization

Many cognitive features are lateralized; that means they predominately involve one hemisphere. Moreover, several brain network disorders show changes in the topology of a region when both hemispheres are compared. A measure to compare both hemispheres is lateralization, which measures a feature in a node of the left hemisphere relative to the same feature in the contralateral node of the right hemisphere. Therefore, for a node, the lateralization index LI, with a potential range from –1 (completely right lateralized) to +1 (completely left lateralized), would be as follows:

$$LI = (F_{left} - F_{right}) / (F_{left} + F_{right})$$

where F is the value of the feature that is observed. If left and right values are similar, so that there is no lateralization, the index would be around 0, with lower index values indicating a lateralization to the right and higher index values indicating a lateralization to the left.

Using Spatial and Topological Features for Network Reconstruction

As for other biological systems, incomplete data sets are a problem for brain connectivity studies. Are there ways to predict whether a connection between two nodes exists? One possibility, tested for the macaque fiber tract and the *C. elegans* neuronal network, is to use local features of a pair of nodes to predict whether they are connected or not (Costa, Kaiser, et al., 2007). Topological features were node degree, clustering coefficient, characteristic path length, and Jaccard coefficient. Spatial or geometrical features included local density of nodes, coefficient of variation of the nearest distances, Cartesian coordinates of the nodes' center of mass, as well as the area size for nodes in the cortical network:

Local density: It is often the case with point distributions (as the centers of mass of the cortical areas) that the number of points per unit area varies along the space. In such cases, it is interesting to consider the *local density* around each point. This value can be estimated by dividing the number P_i of neighboring points contained in a sphere of small radius R centered at the reference point i by the volume of that sphere, that is,

$$L_i(R) = \frac{3}{4} \frac{P_i(R)}{\pi R^3}$$

The quantity $P_i(R)$ has been calculated with respect to each node i by counting how many nodes are at distance smaller or equal to R. Note that this measurement is influenced by the volume of each cortical region. The larger the volume, the smaller the local density.

Coefficient of variation of the nearest distances: Given a reference point i and a maximum radius R, the nearest neighbors Q of that point can be defined as those points which are contained in the sphere of radius R centered at point i. The *nearest distances* of point i are therefore defined as the set of the Euclidean distances between it and each of the nearest neighboring nodes in Q. The coefficient of variation (i.e., the standard deviation divided by the average) of the nearest distances provides an interesting indication about the local distance regularity around each reference point. For instance, a low value of this coefficient indicates that the nearest neighbors of a point are almost equidistant. As with the previous measurement, the coefficient of variation of the nearest distances can also be affected by the volume of the cortical region. More specifically, the larger the volume, the larger this measurement tends to be.

Area size of each cortical region: This measurement corresponds simply to the area size of the two-dimensional surface of each cortical region.

Cartesian coordinates of the cortical areas center of mass: These features, considered together for simplicity's sake, correspond to the x, y, and z coordinates of the center of mass of each cortical area.

Such an approach gave good estimates for reconstructing connections of the macaque visual cortex (Costa, Kaiser, et al., 2007): For the combination of the best two topological and two spatial measures, 111 currently unknown projections were predicted to exist, and 174 connections were predicted to be absent, yielding a realistic ratio for predicted existing connections of 39%, out of all unknown connections. The prediction performance could be further improved through varying the contribution of each feature (Nepusz et al., 2008).

2.4 Dynamic Features

Like for genomics, the hopes are that features of the connectome of a patient can be a biomarker for diseases and an indicator for therapeutic interventions. Identifying biomarkers for diseases based on large-scale genome studies has been challenging. Is the link between connectivity and brain disease also overweighted? What could a structural connectome in principle tell us about the brain organization in health and disease?

In analogy to genetics, we may distinguish a genotype and a phenotype of brain organization. The genotype is given by the structural connectivity observed either at the level of individual synapses (microconnectome) or at the level of fiber tracts between brain regions (macroconnectome) (DeFelipe, 2010). The phenotype represents activity, as seen in fMRI or EEG, or behavior, as for cognitive clinical scores. We refer to these patterns as consequences on dynamics or behavior due to changed brain connectivity (Kaiser, 2013).

The problem of diagnosing a disease, as in genetics, is due to the fact that several mutations of the genotype might result in the same phenotype (disease). Observing brain connectivity, there might be several combinations of changes in fiber tracts leading to hallucinations or seizures, for example. Also, the same connectome organization might lead to different dynamics for changes that affect the internal anatomy and activity of network nodes but not the nodes' topology (see figure 2.10). The idea that many pathways can lead to similar behavior is linked to the concept of degeneracy (Tononi et al., 1999; Price and Friston, 2002), "the ability of elements that are structurally different to perform the same function" (Tononi, Sporns, and Edelman, 1999). If the output (phenotype) is cognitive deficits in patients, the number of connectome (genotype) patterns that lead to such behavior can be seen as the degeneracy of a brain disease. Also, a higher degeneracy, meaning that more connectome patterns are linked to a disease, might result in a higher incidence in a population. A related observation has been made in the field of genetics when linking genetic changes to diseases: multiple genotypes might lead to the same phenotype (heterogeneity) (Addington and Rapoport, 2012). Therefore, detecting one connectivity pattern linked to a disease might only relate to a fraction of all patients. Moreover, many connectome

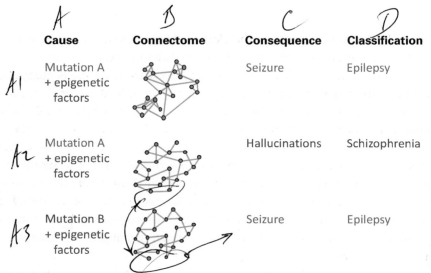

Figure 2.10
Mapping between underlying developmental causes of connectome changes, ranging from genetic factors to spatiotemporal epigenetic factors, to resulting brain connectivity ("connectome"), observable network behavior ("consequence"), and final disease classification. Similar patterns within each of the four categories are shown in red. Note that both genetic patterns and network features alone may be insufficient to inform the clinical diagnosis of a disease: First, the same genetic mutation A might lead to a different connectivity due to different epigenetic factors (first two rows). Second, different genetic mutations A and B could lead to the same connectivity due to additional factors (second and third rows). Third, the same connectivity (second and third rows) might lead to different behavior and disease classification due to changes that solely affect the anatomical organization within individual nodes (adapted from Kaiser, 2013).

changes will be neutral in that they do not lead to a brain disorder; thus, variability in the healthy population is expected to be large as well.

Another problem besides large connectome variability ("noise") is that cognitive deficits might arise from small changes ("signal"). Development can be seen as a system of nonlinear dynamics (Turing, 1952). It has become clear that genetic encoding (Kendler et al., 2011) and self-organization shape the formation of neural systems in health and disease. For self-organization, the interaction with the environment (external factors) or physical constraints (internal factors) can influence the establishment and survival of axonal connections. Consequently, small changes during development might lead to a different connectome and as a result to a different resulting consequence for cognition and behavior of human subjects. As the dynamics in the brain are also nonlinear, a small change in structural connectivity might be sufficient to lead to changes in cognition and behavior. Relatively small changes in connectivity might be sufficient to lead to a brain disorder. Therefore, some connectivity patterns seen in patients might be quite close to the organization of healthy subjects.

To deal with the variability in brain disorders, one approach is the use of computer simulations of brain activity, based on the connectivity in individual patients. Such simulations are already emerging as a way to understand the structural correlates of dynamical changes and disease progression (Deco et al., 2011; Cabral et al., 2012; Raj et al., 2012). As shown above, multiple structural connectivity changes might lead to the same changes in brain dynamics, patient behavior, or clinical test scores. Simulating the activity in the brain of individual patients can inform us about the expected behavioral features and thus about the presence or absence of one subtype of brain disorder. These models can go beyond the observation of patterns in the recordings of brain activity as simulated dynamics could include more complex models. For example, a model based on structural connectivity might include simulated activity of individual neurons or local circuits, which cannot be observed by noninvasive neuroimaging.

3 Evolution of Neural Systems

While organisms adapt to their environment, their solutions are constrained by previous solutions that were found during evolution. For example, the extracellular concentration of salt in humans is the same as in the oceans, despite humans living on land. In addition, for the same problem, solutions may differ. The eyes of mammals and the octopus, while using the same laws of optics for seeing light, have a different anatomical organization. As for other aspects of biology, it is useful to look at brain networks in terms of their evolution (Striedter, 2005). Within the last three decades, we have started to develop pictures of the global connectivity, at varying resolution, of the nervous systems of a number of phylogenetically disparate species. The archetypal "connectome" to be elucidated was that of the hermaphrodite form of the roundworm *Caenorhabditis elegans* (White et al., 1986; Durbin, 1987; Cook et al., 2019), which has a relatively small nervous system allowing for elucidation of a complete wiring diagram. Since then, we have started to discover mesoscale connectomes of the much more complex brains of the pigeon (Shanahan et al., 2013), rat (Burns and Young, 2000), mouse (Oh et al., 2014), cat (Scannell et al., 1995), and rhesus monkey (Felleman and Van Essen, 1991). These species live in different habitats, on water, on land, or airborne, but is this also reflected in a specialized organization of brain connectivity? Even if the functional requirements were similar, did evolution come up with different solutions, as it did for the anatomy of the eye in cephalopods (such as the octopus) and vertebrates?

3.1 Nerve Nets—Cnidaria

There are controversies over which metazoans are the most basal, but according to the conventional view of animal phylogeny, the most ancient metazoans that show neural networks are Cnidaria (such as jellyfish). These animals show a diffuse two-dimensional nerve net in the polyp stage. In terms of network topology, nerve nets represent regular or lattice networks. Such lattice networks, with well-connected neighbors and no long-distance connections, are a fundamental unit of neural systems, existing even in complex systems like the retina and in the layered architecture of cortical and subcortical structures. For Cnidaria, the nerve net receives input from sensory cells, including chemoreceptors, photoreceptors, and tactile receptors, and it innervates motor neurons to generate coordinated behavior. Finally, there is also information about the genetic program for the development of the neural system (Kelava et al., 2015).

3.2 Ganglia—*Caenorhabditis elegans*

Platyhelminths (flatworms) and nemathelminths (roundworms) are the first groups showing a central nervous system. These organisms with a bilateral organization show a body axis with a more or less specialized "head" end. Moreover, there is an aggregation of specialized neurons into ganglia (singular: ganglion), as in the roundworm *C. elegans* (White et al., 1986; Achacoso and Yamamoto, 1992).

The neurons in such ganglia are often not just spatially clustered but also topologically clustered: topological clusters, or modules, are sets of nodes with many connections within a module but few connections between modules. In this way, ganglia can process one modality without interference from neurons processing different kinds of information. Spatial and topological modules do not necessarily overlap, but both tend to be well-connected internally, with fewer connections to the rest of the network. Both spatial proximity and similar topological node features can be good predictors of whether two neurons are connected (Costa, Kaiser, et al., 2007). In addition, similar gene expression patterns tend to occur for pairs of neurons that are interconnected (Kaufman et al., 2006).

In addition to the ganglionic architecture, bilateral nerve fibers run along the body axis of roundworms. For some species, also transverse connections between both fibers exist. Moreover, nematodes show muscle cells that project toward these central fibers in order to get innervated, rather than neurons projecting axons to innervate muscle cells.

The organization into ganglia and nerve fibers is further refined in other animal groups. Annelids show a segmented body plan with bilateral ganglia for each segment while a larger concentration of neurons in the head of the animal emerges through an amalgamation of several anterior ganglia (cephalization). Along with increasing performance of sensory organs, there are first signs of stomatogastric nervous systems that innervate the anterior parts of the intestine.

Arthropods, including insects, show a pronounced concentration of neurons in the head where multiple cerebral ganglia have fused and encompass around 90% of all neurons in the central nervous system. This results in specialized units *within* the fused "brain"—for example, for visual input (the protocerebrum receiving information from the complex eyes and ocelli), for tactile input (the deutocerebrum receiving information from the antennae), and for innervating the head surface (the tritocerebrum).

In mollusks, the central nervous system consists of around five ganglion pairs (cerebral, buccal, pleural, pedal, abdominal). However, this basic arrangement can vary between species through fusion of ganglia or changes in their position. Cephalopods—for example, the octopus—have a complex central nervous system with around 100 million neurons, association regions, and good learning abilities.

3.3 Brains—Vertebrates

Vertebrates exhibit a further "centralization" of central nervous system processing. However, the central processor, the brain, is different from merged ganglia. Brains are distinguished from ganglia by being solely located in the head of an animal while serving the entire body, having two hemispheres with connections (commissures) between them, neurons often forming a surface (gray matter) with fibers forming a core (white matter), and specialized local functions. Due to the further increased cephalization, almost all neurons within the whole nervous system are neither direct sensory nor motor neurons but take part in intermediate processing steps instead.

We also see differences in the features of individual neurons in vertebrates compared to invertebrates. Neurons in vertebrates are more often multipolar (not unipolar), have one axon per neuron (not two or more), have extensive dendritic branching and dendritic spines, have mostly chemical synapses, and have a spontaneous activity around 3–50 Hz (compared to 50–500 Hz).

Starting with chordates, the central nervous system shows a clear distinction into five units: telencephalon, diencephalon, mesencephalon, metencephalon, and myelencephalon. The ventral components of the three latter parts form the brain stem (tegmentum), an ancestral part that differs little between vertebrates. The brain stem includes centers for vegetative function (breathing, heart rate, metabolism, sleep) and motor function (posture) as well as the reticular formation (*formatio reticularis*) a nerve net that receives collateral connections from all fibers that project up from the spinal cord (sensory) or down to the spinal cord (motor). The reticular formation therefore receives multimodal information and also participates in the regulation of the sleep/wake rhythm.

Note that vertebrates also contain ganglia, but they are here defined as nerve cell clusters or groups of nerve cell bodies located in the autonomic nervous system and associated with sensory nerves.

Due to a greater degree of specialization of the brain, we see greater complexity as exemplified by the visual processing system of the rhesus monkey (macaque). Here, the visual module consists of two network components: the dorsal pathway for processing object location and movement and the ventral pathway for processing object features such as color and form (Young, 1992). Such networks, where smaller submodules are nested within modules, are one type of hierarchical network (Kaiser et al., 2010).

As brain size increases, local connections become insufficient for integrating information. Brain networks therefore show a small-world organization that includes not only a high degree of connectivity between neighbors but also long-range connections that act as "shortcuts" linking distant parts of the network. Small-world features are observed in species ranging from *C. elegans* (Watts and Strogatz, 1998) to cat (Scannell et al., 1995), macaque (Hilgetag and Kaiser, 2004), and human (Hagmann et al., 2008), despite different levels of brain size and organization.

In conclusion, the network architecture becomes more complex during evolution, going from a diffuse lattice organization to hierarchical modular networks. Over time, parts of the network specialize, leading to network modules and later to multiple hierarchical levels. There is, however, an associated cost, namely, the protracted period of brain development and functional maturation needed to achieve the specialization. While behavioral traits like parental care provide a buffering mechanism, there remains a wider window of vulnerability, when injury can be harder to recover from (Varier et al., 2011). Overall, evolutionary divergence leads to greater complexity while following essential developmental constraints, like those influencing hub formation, long-distance connections, and modular organization.

The Human Brain, the End of Evolution?

The human and nonhuman primate brain shows several differences from other species, but it is still up to debate which of these changes are unique adaptations and which are a by-product of the increase in absolute brain size (Striedter, 2005). Larger brains tend to have more distinct brain regions, change in brain region proportions, and show a relative decrease in the number of connections, which means a reduced edge density. In turn, moving from the structural connectome of the rat, where more than 50% of all possible fiber tracts exist (Burns and Young, 2000), to the one in primates, where fewer than 15% exist (Felleman and Van Essen, 1991; Young, 1992; Hagmann et al., 2008), necessitates a more modular organization.

There is also a large variation within individuals of the same species. Differences include cortical thickness, size of brain regions, frequency of cell types across cortical layers, brain connectivity, and cortical folding. This large variability, even between healthy individuals, makes the detection and characterization of brain diseases challenging. Given the wide range of environments, which are also changing over time, there is no single optimal phenotype. Instead, a range of behavioral patterns and underlying connectome variations is needed to have a large pool for phenotypic selection. In addition, competition within species and between species leads to separate but equally successful strategies for maximizing fitness (Holmes and Patrick, 2018). For example, individuals might change their choice of food or habitat to avoid strong competition from other individuals of the same species. All connectome solutions have advantages and disadvantages, depending on the environment and the strategies of other individuals; therefore, variation, resulting from genetic or epigenetic factors, can persist within a population.

4 Organization of Neural Systems

In the following sections, species for which information about brain connectivity is available will be presented. Following on from of the roundworm *Caenorhabditis elegans* (White et al., 1986), we have started to discover mesoscale connectomes of the much more complex brains of the fruit fly, pigeon, mouse, rat, ferret, cat, rhesus monkey, marmoset monkey, and human. Given the increasing number of species that are covered, the prospect of comparative connectomics arises (van den Heuvel et al., 2016). All these networks share common wiring principles such as a modular organization, network hubs, and directed links (Kaiser, 2015).

Note, however, that there are several other species for which partial information about the nervous system, usually involving the fibers between the brain and the sensory or motor systems, is available: leech, tadpole at adult (Borisyuk et al., 2008) and larval stage (Ryan et al., 2016), and others. In addition, there is more detailed connectome information for some subsystems of the brain such as the retina in the mouse (Helmstaedter et al., 2013).

4.1 The Worm (*Caenorhabditis elegans*)

Roundworms (nematodes), part of the 12,500 species animal group of the Nemathelminthes, show a constant number of neural cells of each species (eutelic). As their body is not segmented, any movement involves the whole body. As a unique feature of the motor system, muscle cells extend toward the axon to establish a synaptic connection. The roundworm *C. elegans* has been a target of genetic studies showing the links between genes and development and behavior. Its nervous system consists of 302 neurons in the hermaphrodite and 381 neurons in the male form. In addition, the complete cell lineage (Durbin, 1987) and the connectivity of neurons (White et al., 1986) and their spatial positions (Choe et al., 2004) are known (see figure 4.1). *C. elegans* is still the only animal for which the complete connectome is available.

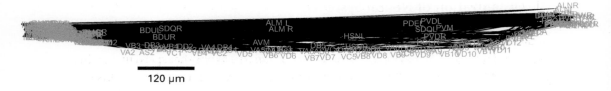

Figure 4.1
Lateral view of the *C. elegans* neuronal network. Note the ventral cord at the bottom and the concentration of neurons in ganglia in the head (left side) (after Kaiser and Hilgetag, 2006).

C. elegans was first described by Emile Maupas in 1900 (Blaxter, 2011). As it is see-through during all stages of development, it is often used for developmental studies. Embryos undergo a stereotypical pattern of cell divisions from the zygote to the larva, such that the cell lineage is mostly invariant (Sulston and Horvitz, 1977; Sulston et al., 1983). We can therefore observe the time course of the formation of the nervous and other systems. There are only 558 cells in the hatching larva and 959 cells in an adult hermaphrodite (excluding the germ line).

The nervous system is characterized by a longitudinal bundle of fibers in the ventral cord of the animal and several ganglia with a higher density of neural cell bodies (White et al., 1986; Hall and Altun, 2008): The diffuse pharynx "ganglion" envelops the pharynx musculature and follows its contours, the anterior and lateral ganglia surround portions of the pharynx muscles and neurons, the small dorsal ganglion lies partially above the lateral ganglion, the ventral ganglion lies below the lateral ganglion, and in the tail some of the small dorsorectal ganglion lies over the anterior portion of the lumbar ganglion (Cherniak, 1994).

While ganglia contain few or no synapses, neuron processes extending from ganglia travel in longitudinal nerve bundles to target regions where they form around 6,400 chemical and 900 electrical synapses (gap junctions). In addition, neurons form 1,500 neuromuscular junctions. Among individual animals, the location of chemical synapses is about 75% reproducible (Durbin, 1987).

4.2 Fruit Fly (*Drosophila melanogaster*)

Insects, originating more than 400 million years ago as part of the 900,000 species animal group of arthropods, show extremities that consist of several units (e.g., legs consisting of multiple cylindrical components). They also show a range of sensory organs to perceive mechanical stimuli, using antennae and other mechanoreceptors, olfactory stimuli, or visual stimuli, using compound eyes and ocelli. Because of their larger body size, up to 624 mm compared to the 1.4 mm of *C. elegans*, insects face the challenge of quickly transmitting information from the head of the animal to the muscles in the legs in order to guide behavior. Unfortunately, myelinated axons that speed up conduction speed by a factor of 10 are undeveloped in invertebrates such as insects. Instead, long-distance connections consist of axons with a larger diameter to increase the conduction speed. Such giant axons, which also appear in the animal group of annelids (e.g., earthworms) make insects a suitable target for electrophysiological studies. At the same time, fruit flies such as *Drosophila melanogaster* are popular for genetic studies because of their small size and their fast generation cycle of 10–12 days.

Drosophila melanogaster's central brain has around 135,000 neurons, many more than the ~300 neurons of the *C. elegans* nervous system, but far fewer than the mouse's 100 million or the macaque's more than 6.3 billion. Shih et al. (2015) established the FlyCircuit database with data from 12,995 projection neurons based on confocal microscopy (see figure 4.2). These projection neurons form 58 bundled neural tracts linking 49 functional units, including 43 local processing units (LPUs), that contain local interneurons with neurites limited within a unit, and six interconnecting units without local interneurons (Chiang et al., 2011). The 49 units form five modules: olfactory, mechano-auditory, left visual, right visual, and premotor. Such a functional specialization is also seen in mammalian brains; for example, the cat brain is composed of visual, auditory, somatosensory-motor, and frontolimbic modules (Scannell et al., 1995).

The *Drosophila* brain also shows a small-world network organization (Shih et al., 2015): a small-world network shows a high connectivity between regions that are connected to some other region (i.e., between neighbors). At the same time, shortcuts, which often connect spatially distant regions (Kaiser and Hilgetag, 2006), ensure that different parts of the network can be reached within few steps (Watts and Strogatz, 1998). Despite their crucial role in speeding up processing in the fly, these connections that often run between modules are usually weaker (contain fewer axons) than many connections at the local level. The importance of these "weak ties" (Granovetter, 1973) was also observed for human functional (Gallos et al., 2012) and macaque structural integration (Goulas et al., 2015). Overall, the *Drosophila* connectome is comparable both to neuronal networks that have been described in *C. elegans* (Watts and Strogatz, 1998) but also to fiber tract networks in the macaque (Hilgetag and Kaiser, 2004), all showing a small-world organization despite different brain size and architecture.

While modules ensure segregated processing of information, the integration of different kinds of information is also needed. Shih et al. (2015) show that, within modules,

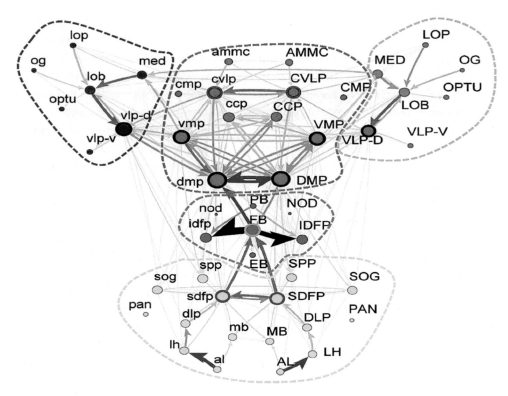

Figure 4.2
Modular structure of the *Drosophila* brain: There are five modules in the brain network: olfaction (yellow), auditory/mechanosensation (magenta), right vision (orange), left vision (purple), and premotor (red). The local processing units (LPUs) with thick black and green boundaries indicate the innermost and second innermost rich-club members, respectively. Node size is proportional to the strength of the LPU. Arrows denote the direction of the connections. Darker and thicker lines represent stronger connections (after Shih et al., 2015).

some nodes have much higher strength than others, potentially functioning as information integrators or broadcasters. These nodes coordinate information flow locally within modules or help to link information of different modules at a more global scale. This local and global integration is comparable to the roles of provincial and connector highly connected nodes (hubs) that have been extensively discussed in other species such as cat, rhesus monkey, and human.

Highly connected nodes, which Shih et al. (2015) measured by the total strength of connections rather than the number of connected nodes (node degree), tend to have stronger connections between each other than would be expected. Such a "rich-club" organization (van den Heuvel and Sporns, 2011), with strong links between well-connected nodes, facilitates synchronization and information integration at the global level. Consequently, removal of these nodes has a relatively severe effect on behavioral performance, consistent with many brain diseases in humans, including Alzheimer's disease and schizophrenia, that have been linked to changes involving rich-club nodes. It remains to be seen what functional consequences rich-club nodes have for *Drosophila*.

As for other networks, in species ranging from *C. elegans* to the macaque, the fly data set contains information about the direction of connections. When direction information is available, we often find that for two connected regions, the connection in one direction might be a lot weaker compared to the opposite direction. In some cases, connections in one direction are absent, leading to one-way streets of information flow. Such asymmetry allows for a larger repertoire of functional circuits with distinct feedforward and feedback loops and might be due to differences in the developmental time windows for synapse formation (Lim and Kaiser, 2015).

Cycles are circuits where information can originate in one node, pass through other nodes, and arrive back at the origin. Using a simulation of information propagation, Shih et al. (2015) show that signals that start within strongly connected loops can persist longer than those that involve weaker loops. Such persistent signals, in this study lasting 10 times as long for strong compared to weak loops, could be crucial for generating stable oscillations or forming memories.

The connectivity of the fruit-fly brain reported by Shih et al. (2015) is based on a reconstruction of a large number of neurons from all brain regions, thereby giving a more complete picture of brain connectivity. Moreover, for many neurons, the secreted neurotransmitter is known, allowing a first estimate of the balance of excitation and inhibition. This balance is not uniform, indicating that local negative feedback is much more crucial for some processing units than for others.

Despite the similarities with the organization of the mammalian brain, it is important to also keep in mind some differences beyond the size of these brains: for the arthropod *Drosophila*, somata of interneurons and motor neurons are commonly unipolar, motor fiber bundles are not myelinated, and they have sensory organs, such as compound eyes, ocelli (simple eyes), and antennae (sensing), which lack obvious counterparts in mammals. Another limitation is that the currently used method is unable to resolve individual synapses. With higher-resolution methods, information about synapses could be gained to estimate synaptic weight, the location of synapses with excitatory versus inhibitory effects in the postsynaptic neuron, and the ability for computation within the dendritic tree of a neuron.

4.3 Pigeon (*Columba livia*)

Birds, in its current form originating more than 150 million years ago, form the 8,800 species animal group Aves. Unlike other vertebrates, they are able to fly, helped by several modifications, including in the nervous system, that are geared toward reducing the weight of their bodies.

Song-learning birds show a particular pattern of cell birth and cell death at the adult stage (Nottebohm, 2005). Seasonal regression of the avian motor pathway leads to a reduction of neurons in the pallial song control nucleus HVC (proper name) between breeding seasons while nearly 68,000 new neurons are incorporated during breeding seasons (Larson et al., 2014). Seasonal network changes are also observed for food-storing birds. The hippocampus, important for retrieval of food cache locations, undergoes change in size and neurogenesis that is correlated with the seasonal pattern in food storing (Sherry and Hoshooley, 2010).

The connectome of fiber tracts between brain regions in the telencephalon of the pigeon (Shanahan et al., 2013) is currently the only bird connectome available (see figure 4.3). It exhibits many of the network properties that have been found in mammals, such as a high small-world index, the prevalence of certain structural motifs, modularity, and the possession of a connective core of hub nodes (Shanahan et al., 2013). Modules, connector hubs, and structural motifs indicate an organization that enables segregation and integration (Sporns and Kötter, 2004; Zamora-Lopez et al., 2010; Sporns, 2013).

The pigeon connectome manifests two levels of modularity with its top-level modules mostly being functionally analogous to those of humans and other primates. While the top-level modules of the human brain are anatomically localized, those of the pigeon brain are more anatomically distributed. The network contains a topologically central connective core, characterized by higher betweenness centrality, higher node degree, and rich-club features. Hub nodes of this core are functionally analogous to hubs in the primate brain (Shanahan et al., 2013). Furthermore, topological modules of the pigeon connectome are functionally and/or anatomically comparable to modules that are revealed when network analysis is carried out on human brains (Hagmann et al., 2008; Shanahan et al., 2013): Both the pigeon and human forebrain possess a module that incorporates prefrontal, premotor, and motor fields and thus links associative sensory with motor areas (Güntürkün, 2005; Hagmann et al., 2008). The pigeon visual module is functionally similar to the human occipital visual module (Shimizu and Bowers, 1999). The viscerolimbic module resembles the human cingulate and paracentral module. Like for humans, the pigeon corticohippocampal module includes areas of the hippocampal complex as well as diverse primary and associative sensory

Organization of Neural Systems

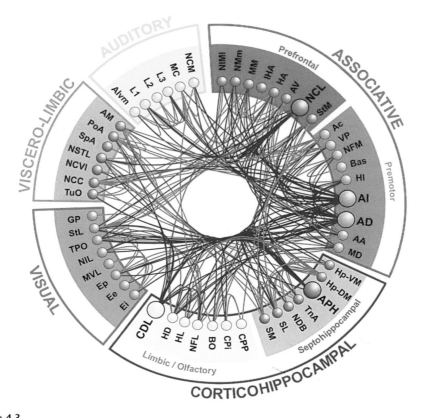

Figure 4.3
The telencephalic connectome of the pigeon forebrain. Network analysis reveals five top-level modules. The associative and corticohippocampal modules can be further decomposed. Connections to and from hub nodes are shown in a slightly darker color (after Shanahan et al., 2013).

systems. The pigeon auditory module, however, does not show a clear correspondence to any module found in humans.

Most of the pigeon's hub nodes are functionally equivalent to hubs in the macaque or cat brain, or to hubs in the human prefrontal and hippocampal formation. In the pigeon brain, three out of five hubs are in the associative module and are of a prefrontal or premotor nature: AD (arcopallium dorsale), AI (arcopallium intermedium), and NCL (nidopallium caudolaterale). The CDL (area corticoidea dorsolateralis) hub, on the other hand, matches connectivity patterns of the cingulate cortex. The fifth hub, APH (area parahippocampalis), resides in the parahippocampal region, a part of the brain that also shows hubs in monkeys and humans.

Overall, despite the absence of cortical layers, the avian brain conforms to the same organizational principles as the mammalian brain concerning its topological network features.

4.4 Mouse (*Mus musculus*)

Mammals, encompassing more than 4,000 species, started a rapid diversification 65 million years ago. For rodents, mice are popular for genetic and pharmacological studies. The mouse brain weighs 0.4 g and contains 75 million neurons. Like for the rat, the mouse brain is lissencephalic, lacking the folded cortical surface of other mammals.

The Allen Mouse Brain Connectivity Atlas project (Oh et al., 2014) and the Mouse Connectome Project reconstruct connectivity of the C57BL/6 mouse brain. The mouse connectome (Zingg et al., 2014) initially reconstructed 240 intracortical connections forming a corticocortical connectivity map that facilitates comparison of connections from different cortical targets. Connectivity matrices were generated to provide an overview of all intracortical connections and subnetwork clusterings (see figure 4.4). The connectivity matrices and cortical map revealed that the entire cortex is organized into four somatic sensorimotor, two medial, and two lateral subnetworks that display unique topologies and can interact through select cortical areas. Together, these data provide a resource that can be used to further investigate cortical networks and their corresponding functions.

A statistical approach (Ypma and Bullmore, 2016), using anterograde viral tract-tracing data provided by the Allen Institute for Brain Sciences (Oh et al., 2014), estimates the connection density of the mouse intrahemispheric cortical network to be 73% while interhemispheric density was estimated to be 59%. The weakest estimable connections (about six orders of magnitude weaker than the strongest connections) could likely represent only one or a few axons.

Moreover, there is detailed information on the connectivity within the mouse hippocampus and between the mouse hippocampus and other brain regions. Including subiculum gene expression patterns revealed a previously hidden laminar organization (Bienkowski et al., 2018).

As for *C. elegans* (Kaufman et al., 2006), similar gene expression is a predictor of connectivity in the mouse (Wolf et al., 2011): the outgoing (incoming) connectivity is successfully predicted for 73% (56%) of brain regions, with an overall fairly marked accuracy level of 0.79 (0.83). Note, however, that this study used *rat* brain connectivity to compare with mouse gene expression as the mouse connectome was not yet available.

In a study of 37 cortical areas and 24 thalamic nuclei using Cre-based labeling, the hierarchical organization of the network only contains two full levels (Harris et al., 2019). Within the hierarchy, most thalamic regions are located at the bottom or top, suggesting that they have pure driver or modulator effects on the cortical areas with which they are connected. For cortical regions, primary visual cortex is at the bottom and the prefrontal area ORBvl (ventrolateral part of the orbital area) is at the top. Moreover, L2/3 and L4 neurons have predominantly feedforward layer projection patterns, whereas L5 and L6 neurons have both feedforward and feedback patterns.

Finally, the distribution of synapse types across different brain regions, the "synaptome," has been mapped out (Zhu et al., 2018). Each brain region showed a distinct fingerprint of synapse types. Areas controlling higher cognitive function contained the greatest synapse diversity, and mutations causing cognitive disorders reorganized synaptome maps. Furthermore, new high-resolution optical methods might bring information about the network at the neuronal, microconnectome level (Li et al., 2010), potentially linking synaptome and connectome information.

4.5 Rat (*Rattus norvegicus*)

The rat brain weighs 2 g and contains 200 million neurons. Despite being larger than mice and less utilized for gene knockout studies (although new techniques could increase the use for gene knockout research), rats are often used for behavioral research involving the memory and reward systems.

Based on early efforts by Burns and Young (2000), 24 regions of the hippocampal formation and associated hippocampus were mapped based on invasive tract-tracing studies. Using cluster analysis, regions linked to "place" and regions linked to "head direction" were identified. Using data from both rat and mouse, the *Brain Architecture Knowledge Management System* contains current information on the rodent macroconnectome (Bota et al., 2012). Unlike for other connectomes, for rats there is detailed information about subcortical structures such as basal ganglia (Swanson et al., 2016) and intrinsic connectivity of globus pallidus (Sadek et al., 2007), hippocampus (Ropireddy and Ascoli, 2011), and amygdala (Schmitt et al., 2012) as well as about cortical structures (Swanson et al., 2018).

Given that the rat connectome has been around 14 years longer than the mouse connectome, there are several interactive software environments for visualizing and analyzing brain connectivity. Tools for rodent connectome data include BrainMaps.org (Mikula et al., 2008), Brain Maps 4.0 (Swanson, 2018), the Neurome Project, and the Allen Institute Brain Atlas (Sunkin et al., 2013). The neuroVIISAS tool (Schmitt and

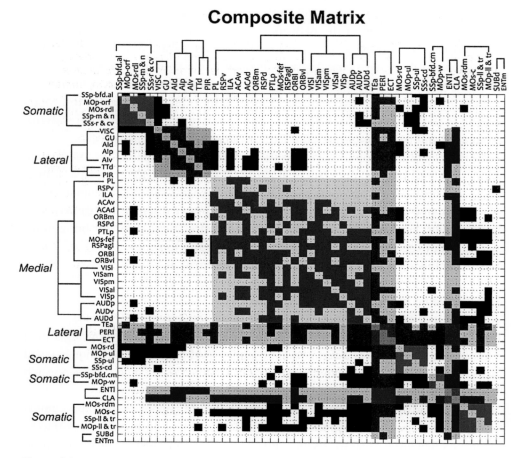

Figure 4.4
Corticocortical mouse connectivity matrix. Combining retrograde and anterograde tracing methods formed the composite matrix, a consensus perspective of corticocortical subnetwork connectivity. Connection origin is listed along the row while targets are listed across the columns. (From Zingg et al., 2014.)

Abbreviations of cortical areas and their subdivisions

ACA: *anterior cingulate area (dorsal, ACAd; ventral, ACAv)*; AI, *agranular insular area (dorsal, AId; ventral, AIv; posterior, AIp)*; AUD, *auditory area (dorsal, AUDd; primary, AUDp; ventral, AUDv)*; CLA, *claustrum*; ECT, *ectorhinal area*; ENT, *entorhinal area (lateral, ENTl; medial, ENTm)*; GU, *gustatory area*; ILA, *infralimbic area*; MOp, *primary somatomotor area (MOp-ll & tr, lower limb and trunk domains; MOp-orf, orofacial domain; MOp-ul, upper limb domain; MOp-w, whisker domain)*; MOs, *secondary somatomotor area (MOs-c, caudal domain; MOs-fef, frontal eye field; MOs-rd, rostrodorsal domain; MOs-rdl, rostrodorsolateral domain; MOs-rdm, rostrodorsomedial domain)*; ORB, *orbitofrontal area (lateral, ORBl; medial, ORBm; ventrolateral, ORBvl)*; PERI, *perirhinal area*; PIR, *piriform cortex*; PL, *prelimbic area*; PTLp, *posterior parietal association area*; RSP, *retrosplenial area (agranular, RSPagl; dorsal, RSPd; ventral, RSPv)*; SUBd, *dorsal subiculum*; SSp, *primary somatosensory area (SSp-ll & tr, lower limb & trunk domains; SSp-m & n, mouth & nose domains; SSp-ul, upper limb domain)*; SSp-bfd, *barrel field of the primary somatosensory area (SSp-bfd.al, anterolateral domain; SSp-bfd.cm, caudomedial domain)*; SSs, *supplemental somatosensory area (SSs-cd, caudodorsal domain; SSs-r & cv, rostral and caudoventral domains)*; TEa, *temporal association area*; TTd, *dorsal taenia tecta area*; VIS, *visual area (anterolateral, VISal; anteromedial, VISam; lateral, VISl; primary, VISp; posteromedial, VISpm)*; VISC, *visceral area.*

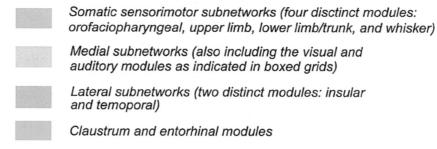

    Somatic sensorimotor subnetworks (four disctinct modules: orofaciopharyngeal, upper limb, lower limb/trunk, and whisker)

    Medial subnetworks (also including the visual and auditory modules as indicated in boxed grids)

    Lateral subnetworks (two distinct modules: insular and temoporal)

    Claustrum and entorhinal modules

Figure 4.4 (continued)

Eipert, 2012) also includes network analysis routines and allows interaction with simulation tools such as NEST (Gewaltig and Diesmann, 2007) to study network dynamics.

Recent analysis of rat connectivity (Swanson et al., 2018) shows that at least 10,000 macroconnections (of a possible 59,292) exist between the 244 gray matter regions identified so far in the right and left cerebral hemispheres of the rat (122 corresponding regions in each side). Moreover, using multiresolution consensus clustering for a hierarchical network analysis shows four subsystems at the top level (see figure 4.5). Finally, further analysis indicates that the status of a region as a connectivity hub depends on the size and coverage of its anatomical neighborhood.

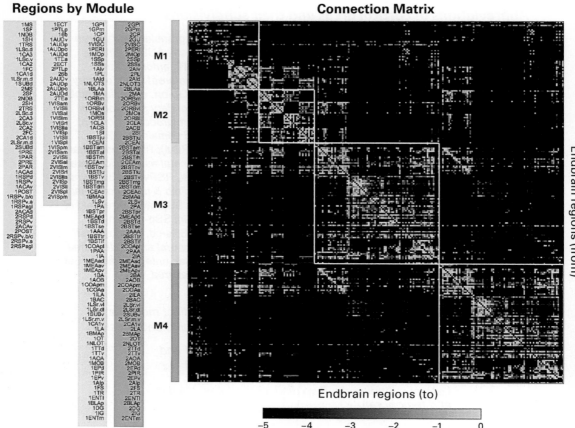

Figure 4.5
Endbrain connection matrix (both hemispheres) with connection weights displayed on a $\log_{10}$ scale and arranged in an ordering that matches the hierarchical ordering delivered by multiresolution consensus clustering (MRCC). MRCC yields four top-level modules (subsystems, clusters) for the complete set of 244 (122 per hemisphere) endbrain regions. The prefix "1" or "2" for region names refers (respectively) to "side 1" (left or right) or "side 2" (left or right) of the endbrain (after Swanson et al., 2018).

There is also detailed information about other parts of the forebrain such as the interbrain, consisting of thalamus and hypothalamus. Compared to cerebral cortex, cerebral nuclei, and the hypothalamus, the 46 thalamic nuclei of the rat are sparsely connected with an edge density of 17% within and of 5% between hemispheres (Swanson et al., 2019). Only one region, the reticular thalamic nucleus, forms a hub. For the hypothalamus, with 65 regions on each side of the brain, two network subsystems deal with the control of physiological and behavioral function (Hahn et al., 2019). Furthermore, there is a correlation between node dominance, as measured through centrality measures (degree, strength, betweenness, and closeness), and an inhibitory role, as measured through gene markers (GAD65).

4.6 Ferret (*Mustela putorius furo*)

The ferret, the domesticated form of the European polecat, has a brain that weighs 3.1 g and contains 400 million neurons. Ferrets are part of the oldest recent group of carnivores with males being substantially larger than females. As they feed on much smaller prey, leading to less available energy, their brains contain fewer neurons and have a lower neural density than would be expected given their cortical mass (Jardim-Messeder et al., 2017).

Compared to other model systems, ferrets have a distinct advantage for developmental studies: because of slower brain development, many processes that occur before birth in other species, such as cortical folding (gyrification), mainly occur later on in ferrets. This also allows observation of the effect of postnatal interventions—for example, lesions—on the ongoing brain development (Sukhinin et al., 2016). Furthermore, ferrets share substantial homologies with another species for which a connectome is available, the cat (Manger et al., 2010). Finally, electrophysiological studies have linked neural dynamics to behavior, leading to a model system that allows the anatomical and functional study of early brain development.

Concerning the ferret connectome, the Ferretome database contains a web interface to search for tract-tracing and cytoarchitectural data. Based on 150 studies, the database contains 20 unique injection sites with 200 labeling sites as well as cytoarchitectural data for 12 areas, primarily for visual and auditory cortex (Sukhinin et al., 2016).

4.7 Cat (*Felis silvestris catus*)

The domestic cat has a brain that weights 25 to 30 g, about 1% of the body mass, and contains around 1.2 billion neurons (Jardim-Messeder et al., 2017). Cats not only have an advanced visual system but can also suffer from "human" conditions such as cognitive decline and dementia.

The cat macroconnectome, considering cytoarchitecture and physiology for cortical parcellation, incorporates data of one hemisphere with 65 regions and 1,139 projections between them (Scannell et al., 1995; Scannell et al., 1999). The strength of a connection is measured on an ordinal scale with 1 reflecting a weak connection, 2 reflecting a medium strength connection, and 3 reflecting a strong connection. A search for topological modules (Hilgetag, Burns, et al., 2000) identifies four modules corresponding to visual, auditory, frontolimbic, and somatosensory-motor function (see figure 4.6).

The cat connectome also shows a rich-club organization (de Reus and van den Heuvel, 2013) with hubs mostly at the boundaries between modules. Furthermore, 86% of the connections between modules consisted of rich-club connections, linking rich-club nodes, and feeder connections, linking non-rich-club nodes to rich-club nodes.

This global connectome organization is also linked to local features within brain regions: Relative cytoarchitectonic differentiation, given the thickness and neural density of cortical layers, as well as spatial relations, given by the border distance between brain regions (adjacent, next-neighbor-but-one, etc.), are good predictors of cortical connectivity in the cat brain (Beul et al., 2015).

At the microconnectome level, the connectivity between different neuronal types of the primary visual cortex, area 17, has been studied extensively (Binzegger et al., 2004). Looking at individual neurons, using the box-counting method on 3-D reconstructions of axonal trees yields a fractal dimension of around 1.5 across different types of neurons (Binzegger et al., 2005).

Information on connectivity within regions and between regions has also been used for simulation of neural dynamics (Zhou et al., 2007). For a model of area 17 (Binzegger et al., 2009), given a reasonable choice of the ratio between excitatory and inhibitory efficacy, the overall cortical circuit lies near the border of dynamical stability. For a dynamical simulation of the network between brain regions (Zhou et al., 2006), the dynamics exhibit a hierarchical modular organization revealing functional clusters coinciding with the anatomical communities described above.

There are also some structural connectivity studies using DTI. Concerning early development, in newborns the main body of the thalamocortical tract was smooth, and fibers branching from it were almost straight, while the main body became more complex and branching fibers became curved, reflecting gyrification, in the older cats. Corticocortical tracts in the temporal lobe were smooth in newborns, and they formed a sharper angle in the later stages of development. The cingulum bundle and superior

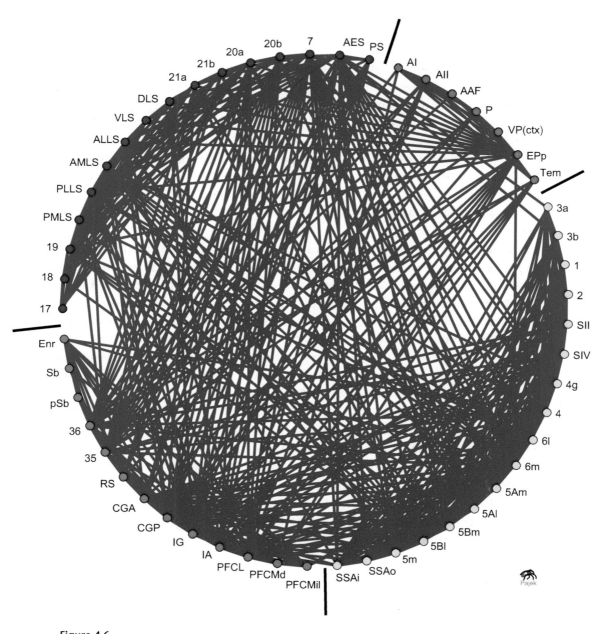

Figure 4.6
Clustered structure of cat corticocortical connectivity. Bars indicate borders between nodes in separate clusters. Cortical areas were arranged around a circle by evolutionary optimization, so that highly interlinked areas were placed close to each other. The ordering agrees with the functional and anatomical similarity of visual (blue), auditory (red), somatosensory-motor (yellow), and frontolimbic (green) cortices (adapted from Hilgetag and Kaiser, 2004).

longitudinal fasciculus became more visible with time. Within the first month after birth, structural changes occurred in these tracts that coincided with the formation of the gyri (Takahashi et al., 2010). Concerning late development, cats as well as dogs can show cognitive decline, cognitive dysfunction syndrome, and dementia, including β-amyloid plaques and neurofibrillary tangles, leading to potential animal models for Alzheimer's disease. It will therefore be interesting to see future diffusion imaging studies across the life span of the cat.

4.8 Rhesus Monkey (*Macaca mulatta*)

Rhesus monkeys, a species of macaques, are omnivores and old-world monkeys, native to northern India, Myanmar (Burma), Southeast Asia, and eastern China. Their brain weighs 87 g and contains 6.3 billion neurons (Herculano-Houzel et al., 2007). Monkeys have been used for studies of the neural mechanisms of cognition as well as for clinical applications such as developing the polio vaccine or developing deep-brain stimulation. Most of this work has been carried out on macaque monkeys, although there are also studies on marmosets, and a few studies on squirrel monkeys and capuchin monkeys. There have been no such studies on great apes such as the chimpanzee or the lesser apes, the gibbons (Passingham, 2009).

In 1991, the first overview of macaque cortical connectivity was published with a focus on the visual system (Felleman and Van Essen, 1991). Connections, based on dozens of individual tract-tracing studies, were reported on an ordinal scale with 1 for weak, 2 for intermediate, and 3 for strong connections. Including additional studies led to the development of CoCoMac, the first online database of connectome data, in 1996 (Stephan et al., 2001; Kötter, 2004; Stephan, 2013). A systematic study of projections between 29 regions in the same hemisphere and to some regions in the contralateral hemisphere has recently been reported by Markov and colleagues (Markov, Ercsey-Ravasz, Lamy, et al., 2013; Markov, Ercsey-Ravasz, Van Essen, et al., 2013; Markov et al., 2014).

A topological cluster analysis of the visual system of the macaque connectome indicated a separation into two streams, which were identical to the two functional visual streams: the ventral stream for object recognition and the dorsal stream for object position and movement (Young, 1992). In particular, further analysis identified two topological modules, occipitoparietal and parietal, and inferior-temporal and prefrontal, while the primary visual cortex, area V1, is separated from both clusters (Hilgetag, Burns, et al., 2000).

Networks seem to maximize their dynamic repertoire. While network motifs (Milo et al., 2002) can be seen as the underlying information infrastructure of a network ("structural" motifs), at each time only a subset of connections, connecting all nodes of a structural motif, will be used ("functional" motif). The macaque structural connectome maximizes both the number and the diversity of functional motifs, while the repertoire of structural motifs remains small, providing support for both functional segregation and functional integration (Sporns and Kötter, 2004).

Given the center of mass of brain regions, the spatial features of networks can be evaluated as well. As fiber trajectories were not available within the traditional invasive tract-tracing data sets, the lengths of fiber tracts are estimated by the Euclidean distance between connected regions. In other words, tracts are treated as straight lines between network nodes. Within a hemisphere, most connections project toward adjacent or nearby cortical regions. However, 10% of the connections link areas that are further than 4 cm apart (Kaiser and Hilgetag, 2004a). Long-distance connections existed, for example, between areas 10o and 7a, V2 and 46, and V3 and 46. Some connections span almost the whole length of the hemisphere.

The spatial distance between nodes can also be employed as a network communication strategy. Instead of using shortest paths, information can pass through the network by activating the (spatially) nearest node. With this measure of navigability, mammalian cortical networks, including the macaque connectome, were found to show near-optimal communication. Efficiency of communication was comparable to that of using a shortest-paths approach (Seguin et al., 2018).

Finally, neuroimaging-based information about macaque connectivity is also becoming available. The PRIME database, an initiative of 22 nonhuman primate imaging sites, is collecting structural scans and DTI data (Milham et al., 2018). This initiative also aims to include other species, such as marmosets, in the future.

4.9 Marmoset Monkey (*Callithrix jacchus*)

Marmosets are more accessible for some experimental procedures because of their lissencephalic brain, twin births being the norm, and the development of transgenic animals (Lin, Takahashi, et al., 2019). Marmosets have a smaller brain ($\approx$ 35 mm × 25 mm × 20 mm), comparable in size to some rodents (e.g., squirrels), with a brain mass and number of neurons only 10% of that of macaques (Herculano-Houzel et al., 2007).

New World monkeys such as marmosets evolved in isolation from Old World monkeys (including macaques), apes, and humans for at least 40 million years (Lin, Takahashi, et al., 2019). However, they show rich social behavior and vocal communication.

For the common marmoset (*Callithrix jacchus*), naturally found in South America, a database of tracer injection studies in about 50 cortical areas is available online (Majka et al., 2016). Current studies indicate that visual, auditory, and motor systems of the marmoset are very similar to those of the macaque (Bakola et al., 2015). Figure 4.7 shows a connectivity matrix between all regions where fiber tract information about both directions, using anterograde and retrograde tracers, was available.

4.10 Human (*Homo sapiens*)

The brain of full-term human newborns weighs 350 g, around 1,000 g one year after birth, and 1,300 g at puberty. The adult human brains contain 16 billion neurons and weighs around 1,500 g. While information from other species has been around for a long time, data about human neuroanatomy was scarce (Crick and Jones, 1993). This led to proposals in 2005 to determine the human connectome in terms of connectivity between brain regions but also connectivity between neurons (Sporns et al., 2005).

Using noninvasive neuroimaging techniques, a first macroconnectome already became available in 2008: By using DSI, Hagmann and colleagues (Hagmann et al., 2008) noninvasively mapped structural connectivity within and across cortical hemispheres in individual human participants (see figure 4.8). A structural core within posterior medial and parietal cerebral cortex is characterized by nodes with high degree, high connection strengths, and high betweenness centrality. The structural core contains brain regions that form the posterior components of the human default mode network and acts as connector hubs that link all major structural modules. Both within and outside of core regions, there is a high correspondence between structural connectivity and resting-state functional connectivity measured in the same participants (Hagmann et al., 2008).

These early results (Hagmann et al., 2008) were based on five subjects, but as brain connectivity of an individual is as unique as a fingerprint, larger data sets are essential to understand the relation between network structure and network function. The first large-scale endeavor to measure the human connectome at the population level was the Human Connectome Project (Van Essen et al., 2012). Using a standardized approach for generating structural connectivity (Glasser et al., 2013), the Human Connectome Project assembled data from 1,200 healthy participants—including adults as well as young adult twins and nontwin siblings—to provide a baseline for human brain network organization. Inspired by the Human Connectome Project, several other projects have commenced, including the Developing Human Connectome Project (Fitzgibbon et al., 2016) and disease-related projects (see part III for further information).

An initiative at an even larger scale is the UK Biobank Imaging project (Miller et al., 2016). The UK Biobank assembled data from 500,000 UK residents, including blood

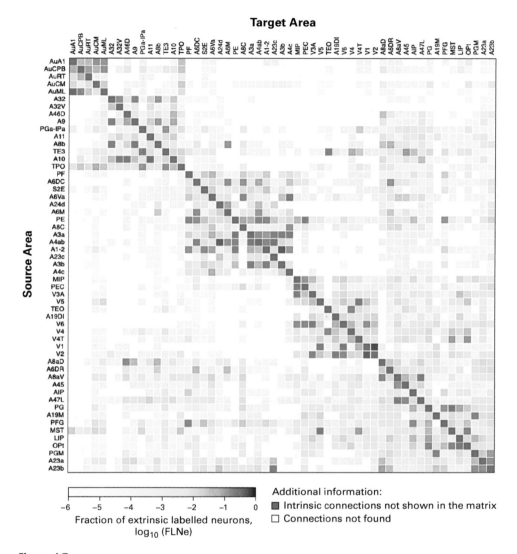

Figure 4.7
Marmoset structural connectivity. Weighed and directed connectivity matrix based on the results of injections of monosynaptic retrograde fluorescent tracer injections. The rows and columns of the matrix represent individual cortical areas with their order following hierarchical clustering. The Targets are the injected areas while the Sources indicate the areas in which the projections originate. This view was generated from http://analytics.marmosetbrain.org/ with injection data described in Majka et al. (2016); Lin, Takahashi, et al. (2019).

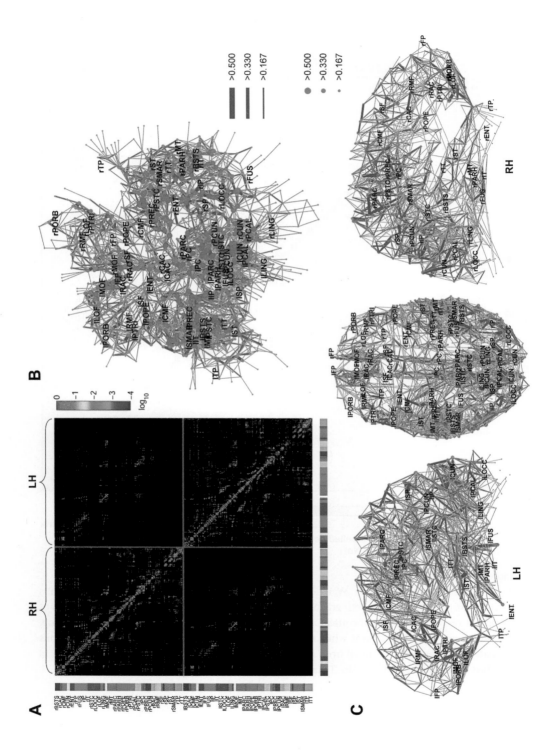

Figure 4.8

High-resolution human connectome matrix, network layout and connectivity backbone in one subject. (A) Matrix of fiber densities (connection weights) between all pairs of n = 998 regions of interest (ROIs). ROIs are plotted by cerebral hemispheres, with right-hemispheric ROIs in the upper-left quadrant, left-hemispheric ROIs in the lower-right quadrant, and interhemispheric connections in the upper-right and lower-left quadrants. All connections are symmetrical and displayed with a logarithmic color map. (B) Kamada-Kawai force-spring layout of the connectivity backbone. Labels indicating anatomical subregions are placed at their respective centers of mass. Nodes (individual ROIs) are coded according to strength, and edges are coded according to connection weight (see legend). (C) Dorsal and lateral views of the connectivity backbone. Node and edge coding as in (B). (From Hagmann et al., 2008.)

---

samples, genetic information, socioeconomic data, health care records, and cognitive skills assessments. Of these subjects, 100,000 are undergoing MRI whole-body scans, including brain scans for structural (diffusion imaging) and functional (rs-fMRI) connectivity (Alfaro-Almagro et al., 2018). Importantly, this data set will also include longitudinal data as dementia patients will be rescanned. This will not only allow the observation of disease progression but will also allow for the development of early biomarkers as many dementia patients were symptom-free at the time of their initial UK Biobank scan.

II Connectome Maturation

Now that I have described, in part I of this book, the spatial and topological features of the connectome, in part II we will observe how these features change during maturation, from conception until adulthood. In particular, I will discuss how mathematical models and computer simulations can be used to understand mechanisms for connectome development. Computer models allow us to test different hypotheses of how network architectures can arise and to better understand the interaction of different mechanisms.

While several models are at an abstract level, including only a small subset of biological factors, it is increasingly possible to compare their outcomes with biological data. Large databases include information about neuron morphology, about connectivity between neurons and brain regions, and about changes in connectome organization both before and after birth. At the same time, several computational tools for modeling development of individual neurons, of small networks, and of brain tissue are available. All these developments bring us closer to the ultimate aim of predicting the outcome of perturbations occurring during brain development, understanding the underlying changes that cause brain disorders, and using this information to inform therapeutic interventions (Silva et al., 2019).

Note that computational modeling of brain development is a wide field. Within this book, we will focus on mechanisms that are linked to the establishment of connections, layers, and cortical folding. For other patterns, even though some will be briefly mentioned, readers are encouraged to consult the literature concerning the formation of retinal mosaics (Eglen et al., 2000), ocular dominance columns (Rakic, 1986), and the development of the morphology of neurons (Koene et al., 2009). An overview of neural development models, complementing the ones mentioned in this book, can be found in van Ooyen (2003) and van Ooyen (2011).

5 Brain Development

5.1 Overview

Although the modeling of connectome changes involves changing connections between nodes, it is useful to know the rudiments of brain development before neural connections have been made to look at early pattern formation (van Ooyen, 2011). The fertilized egg undergoes several cell divisions leading to the formation of multiple layers during gastrulation. At this stage, the inner layer (endoderm), the intermediate layer (mesoderm), and the outer layer (ectoderm) can be distinguished. The mesoderm, along with cartilage and bones, forms muscles and the notochord which later becomes the vertebral column. The ectoderm, along with the epidermis, forms the neural plate—the basis of the nervous system. During neurulation, the edges of the neural plate fuse together forming the neural tube later developing into the central nervous system. In addition, neural crest cells from the epidermis start to form the peripheral nervous system. For the neural tube, as a result of molecular gradients, a number of distinct domains arise which are the precursors of different areas (regionalization). Neurons are born (cell proliferation) and move outward toward the surface of the neural tube (cell migration). While migrating, the process of forming different neural and glia cell types starts (cell differentiation). This early stage of proliferation and migration also shows a large amount of cell death (apoptosis).

Subsequently, connections between neurons are formed. Observing the development of a connectome at these early stages poses many challenges as this cannot be done directly. While one may look at the growth of an individual axon under the microscope, the growth of all cells cannot be observed simultaneously. Our views of connectomes are only snapshots at different stages of development and, except for the Developing Human Connectome Project in humans (Fitzgibbon et al., 2016), usually involve only the adult stage. In addition, observing the growth of axons is an invasive procedure that either involves *ex vivo* neural cell cultures of neurons or resected brain tissue. Despite these limitations, there are several types of evidence that might elucidate how neuronal networks originated during development even though we cannot directly observe what happened (Collin and van den Heuvel, 2013; Di Martino et al., 2014).

5.2 Patterns of Brain Regionalization

Ocular Dominance Pattern

Pattern formation occurs in many systems, including the somatosensory, the auditory, and the olfactory systems; however, I will use the visual system as an example here. Within the primary visual cortex (Brodmann area 17), ocular dominance patterns are formed during early development. In monkey striate cortex, for example, neurons become preferentially activated by one eye—one eye being dominant above the other eye (Hubel, Wiesel, and LeVay, 1977; Hubel and Freeman, 1977). Cells then form a spatial pattern with distinct regions for one eye and other regions for the other eye. These groups of cells have been characterized by their receptive fields through electrophysiological recordings and by their projection patterns: labeled material injected in one eye can travel to the visual cortex and then be visible in autoradiography (Hubel, Wiesel, and Stryker, 1977) and silver staining of tangentially running fibers in layer IVc (see figure 5.1).

Topographic Mapping

As an example of patterns between regions, the fiber tract between areas can follow a particular ordered organization. For a region, different features of the outside world can be represented on its surface in different locations. For example, the somatosensory map represents different neurons that are linked to different receptors in the skin observing touch while the visual map in the primary visual cortex represents different inputs from the retina. The process of connecting neurons in one region to neurons in another region can be seen as a mapping function. In order to preserve the spatial representation of the outside world, when information is transferred from one region to another, neighbors of a neuron in one region need to have a similar response to stimuli in the outside world (the receptive field) to that of neighbors of a connected neuron in the other region; that means the mapping has to be topographic. A topographic map between two brain regions A and B is a map where a representation that is adjacent in one map is also adjacent in the other map. Note that this definition allows for scaling,

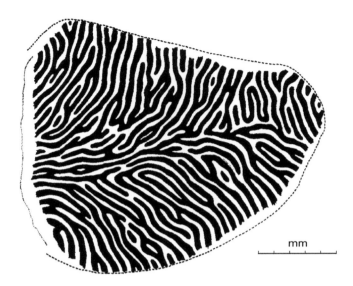

Figure 5.1
Ocular dominance pattern: Reconstruction of the ocular dominance columns in area 17 of the right hemisphere of a rhesus monkey (tangential section, lateral is to the right, anterior is up, and the horizontal bar represents 5 mm). In the diagram every other column has been inked in; thus the dark stripes in the figure correspond to one eye and the light stripes to the other (after Hubel and Freeman, 1977).

representation for a feature being enlarged in one map, and flipping along the axis (see figure 5.2).

Refinement of Synaptic Connectivity

Distinct connectivity between entities also occurs in the peripheral nervous system when motoneurons connect to muscle fibers. In muscle fibers of vertebrates, each fiber is innervated by one motor neuron at the mature stage. During early development, however, each fiber receives synaptic connections from 5 to 10 different motor neurons. As there is little, if any, motor neuron death, this change must be due to the retraction of axons and removal of synapses during development (Willshaw, 1981).

This initial "superinnervation" is also seen in other parts of the nervous system where cells are commonly innervated by more axons than they ultimately maintain into adulthood (Purves and Lichtman, 1980; Rakic et al., 1986; LaMantia and Rakic, 1990; Lohof et al., 1996; van Ooyen, 2001). This occurs, for example, in the innervation of sympathetic and parasympathetic ganglion cells (Purves and Lichtman, 1980),

Positions in brain region A	Positions in brain region B	Topographic mapping?
	3 2 1	
	6 5 4	Yes (flipping, here along the x-axis)
1 2 3	9 8 7	
4 5 6		
7 8 9	3 2 1	
	6 5 4	Yes (scaling, here logarithmic)
	9 8 7	
	1 2 9	
	4 5 6	No (neighborhood relation destroyed)
	7 8 3	

Figure 5.2
Topographic mapping. A map in one region, brain region A, and three potential projections to another region, brain region B. Neurons with the same number are connected. The topographic neighborhood relation is preserved for flipped or scaled maps but not for the shuffled map at the bottom (8 and 3 are neighbors in that map but not in the original map).

the formation of ocular dominance layers and columns in the macaque lateral geniculate nucleus (LGN) and visual cortex (e.g., Wiesel, 1982), and the climbing fiber innervation of Purkinje cells in the cerebellum (Crepel, 1982).

5.3 Mechanisms for Pattern Formation

Neuromorphic Fields

For pattern formation in 2-D or 3-D, neuromorphic fields can be an important concept. Proposed by Alan Turing in 1952 (Turing, 1952), molecules can act as "morphogens" by diffusing and interacting with each other in space. In most models, the substrate for diffusion is assumed to be a two-dimensional space, but similar models can be constructed for three dimensions if needed. These models are also called reaction-diffusion models. While they can be used for a wide range of biological phenomena—for example, the skin and fur pattern of tigers (Murray, 2003)—such models have also been applied to look at pattern formation within brain development. Note that pattern formation here does not necessarily refer to network topology but to factors that define the identity of neurons which may or may not relate to distinct connectivity patterns.

Each model consists of at least two molecules (morphogens), different diffusion rates, and specific interactions between morphogens. If morphogens only vary over time, their concentration can be described by the equations

$dU/dt = R_1(U,V),$
$dV/dt = R_2(U,V),$

with R_1 and R_2 being nonlinear functions that control the production rate of U and V.

In biological systems, the concentration of the morphogen would depend not only on time but also on the spatial position, due to diffusion of molecules. In that case, modified equations would include diffusion coefficients that control the spreading rate over space.

One particular example of a reaction-diffusion system is the Gierer-Meinhardt model (Gierer and Meinhardt, 1972; Meinhardt, 2008) where one morphogen is an activator promoting synthesis of itself and the other morphogen is an inhibitor that inhibits its own growth but also that of the activator morphogen. Depending on the model parameters, a stable spatial pattern can emerge over time where the concentration of each morphogen at a certain location does not change anymore. Note that there is an additional model parameter outside the equations: the size of the embedding space in which the morphogenetic field is generated. The field size can influence the pattern that will emerge after stabilization. As a consequence, pattern formation early on, when a neural system is small, will differ from the outcome of a pattern formation process that starts later on, when the embedding space is larger (Murray, 2003).

Morphogenetic fields have been discussed in a wide range of developmental phenomena including cortical folding (Toro and Burnod, 2005), the parcellation of the cortical surface into different brain areas (Mallamaci and Stoykova, 2006), and axes formation in the retina (Picker et al., 2009).

An example of reaction-diffusion systems for pattern formation is the formation of retinal mosaics (Sernagor et al., 2006). Ganglion cells of each type are distributed regularly over the retina, forming regular hexagonal patterns (mosaics). It has been proposed that cells follow a minimum spacing rule where new cells locate randomly but keep a minimum distance to any already existing cell. Mathematical models of mosaic production could reproduce the patterns in chick ganglion cells. One proposed model uses lateral inhibition with a reaction-diffusion system with Delta and Notch molecules (Eglen and Willshaw, 2002). In such a model, Delta encourages a cell to form into a ganglion cell and Notch drives it to mature into another cell type. The competitive mechanism is that a higher concentration of Delta in a cell will lead to an increased concentration of Notch in neighboring cells, which in turn reduces the concentration of Delta in these neighbors. In other words, one cell becoming a ganglion cell will reduce the likelihood that neighboring cells also become ganglion cells. As a result, ganglion cells will be regularly spaced out across the retina. However, numerical simulations have shown that this lateral inhibition in itself is insufficient to form mosaics and that cell death, as an additional model component, is needed (Eglen and Willshaw, 2002).

Biologically, the production of morphogens can be seen as a gene expression pattern where expression can be influenced by internal and external factors. For example, gradients of chemical molecules in the external space can result in genes being expressed only in distinct spatial locations, where the concentration of these molecules is above a certain threshold. Such gene regulatory models compare well with experimental observations (Meinhardt, 2008).

For neural systems, gene expression across the brain has been observed in many different species. Gene expression in different brain regions was measured in human postmortem studies (see figure 5.3). Broad spatial gradients were conserved in different species and are linked to regional cellular architecture, microcircuits, and connectivity between regions (Fornito et al., 2019).

Competition

Molecular factor competition

Retraction of connections as a refinement of neuronal connectivity can be seen in many different systems, from the neuromuscular junction leading to each muscle fiber being innervated by only one motoneuron, to other systems. One general modeling idea is that axons from different neurons compete to increase the synaptic strength of their connection(s) with the target cell (for a more comprehensive overview of developmental models, see chapter 10 of Sterratt et al., 2011). At the same time, another process is involved to remove axons with weaker synaptic strength. As a result, the number of neurons connected to the target neuron is reduced, leading, for the case of the neuromuscular junction, to only one neuron innervating a muscle fiber.

In an early model (Willshaw, 1981), there is competition between two factors: a factor at the muscle end plate that degrades all synapses and another factor that strengthens synapses so that the total synaptic strength of all synapses remains unchanged. The change in synaptic strength S_{nm} for one particular synapse, connecting a motoneuron n with a muscle fiber m, can be calculated as follows:

$$\frac{dS_{nm}}{dt} = -\alpha M_m + \beta S_{nm}$$

Brain Development

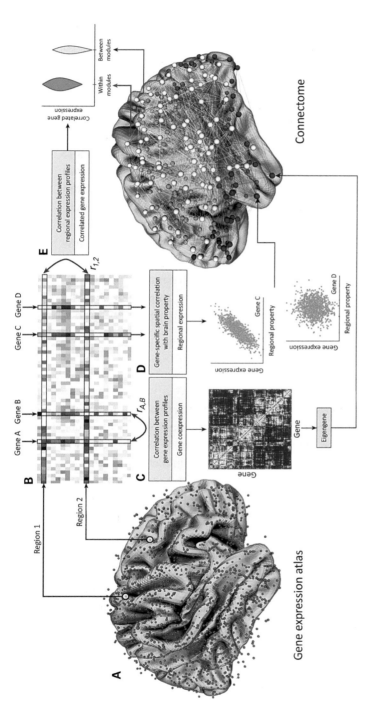

Trends in Cognitive Sciences

Figure 5.3

Workflows for relating transcriptomic atlases to connectome data. (A) Brain surfaces show the locations of different cortical tissue samples taken from the six donor brains comprising the Allen Human Brain Atlas (left) and an example connectome constructed using diffusion magnetic resonance imaging data (right), where colors indicate node membership to different topological modules. (B) Each sample in the expression atlas is characterized by a vector of expression values across genes. These region-specific vectors can be combined into a region × gene expression matrix. (C) Gene co-expression is the correlation between the regional expression profiles of pairs of genes (i.e., columns of the expression matrix). It can be estimated for each pair of genes to obtain a gene co-expression matrix. Summary measures of expression patterns in this matrix, such as eigengenes, can then be mapped back onto the brain and related to regional properties of network organization. (D) Analysis of regional expression patterns involves extracting the spatial profile of each gene separately and correlating it with regional variations in some brain property (e.g., volume, activation, or node degree). In this hypothetical example, Gene C shows a spatial correlation with the brain property whereas Gene D does not. (E) In a correlated gene expression (CGE) analysis, vectors of gene expression for each region (rows of the expression matrix) are correlated with each other. CGE values can then be related to pairwise measures of brain structure or function. In this hypothetical example, the distribution of CGE values for connections within modules is compared to CGE values for connections between modules (after Fornito et al., 2019).

where M_m is the mean synaptic strength at end plate m. The parameters α and β determine the relative influence of synapse degeneration and synapse strengthening, respectively. Both parameters α and β are adjusted so that the total amount of synaptic strength per motoneuron remained unchanged. This model can reproduce the change from initial "superinnervation" to innervation by a single motoneuron but also the experimentally observed pattern that large motor units lose more connections than smaller ones.

An alternative model uses competition at the end plate for a limited resource that is initially assigned to each muscle fiber (Gouzé et al., 1983). Larger synapses take up more resources than smaller ones, so synapses that arrive or have a larger initial size will consume more resources and leave fewer resources to other competing synapses. Once all resources are depleted, weaker synapses will be removed, leading, depending on the threshold for removal, to single innervation. In addition to including only one molecular factor, the resource at the end plate, this model also includes electrical activity of each motoneuron. It is one example of mixed models with neurotrophic factors and neuronal electrical activity (van Ooyen and Willshaw, 1999). In general, depending on the neural system that is studied, such a combination of factors might be needed to reproduce the observed wiring formation.

Activity-Dependent Competition

Lateral inhibition

While we already saw neuronal activity as a factor in molecular models of synaptic strength changes, there is a special case for the use of inhibition. Two neurons will have a negative interaction and reduce synaptic density or synaptic strength of other neurons within a sheet of neurons. An example of where such models can be used is the formation of ocular dominance columns (Swindale, 1980; Huberman, 2007). The visual cortex is represented by a two-dimensional surface with neurons representing the neurons of cortical layer IVc. Connections from each eye, with an intermediate stop at the LGN, would reach the visual cortex. Synapses of one eye would have a positive influence on nearby synapses of the same eye but a negative influence on further-away synapses of the same eye. In contrast, they would have a negative influence on nearby synapses of the opposite eye and a positive influence on further-away synapses of the opposite eye. Note that there would be no influence in either case for synapses that are even further away. The distance-dependent interaction between synapses from the same eye can be modeled as a Mexican hat (also called difference of Gaussians) function (see figure 5.4) while the interactions between synapses from different eyes follow an inverse Mexican hat function.

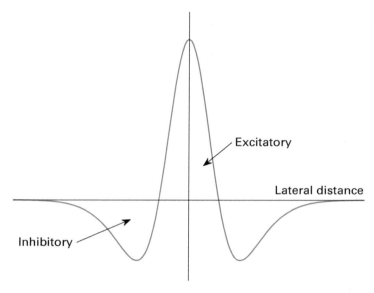

Figure 5.4
Mexican hat function. Interactions with neurons in the spatial vicinity, nearby on the two-dimensional surface, of a neuron are positive (excitatory) while interactions further away are negative (inhibitory).

The model starts with a random allocation of left- and right-eye synapses for neurons in the visual cortex. Following competition, the patterning into dominance columns as observed experimentally occurs (cf. figure 5.1). The width of the dominance stripes depends on the width of the Mexican hat interaction function. Note that interactions in this model rely on changes in synapse densities without providing a specific biological mechanism for these changes.

An earlier model has indicated neural activity as a potential source for this competitive mechanism (von der Malsburg and Willshaw, 1976). In that model, synapses from the same eye show correlated firing whereas activity from synapses of different eyes are anticorrelated. Due to Hebbian learning, interactions between same-eye cells are then strengthened and between different-eye cells are weakened.

Structured input, where nearby cells are active around the same time, is a crucial component of many models—for example, spontaneous retinal waves during early development of the retinotectal tract (Hennig et al., 2009)—but might not be needed in all cases. A recurrent neural network of binary threshold neurons with initial random wiring can form neural assemblies based on Hebbian learning. This model also predicts an ongoing reorganization of assemblies, even at the mature state, where the number of assemblies remains stable but neurons can change their membership (Triplett et al., 2018).

Topographic mapping

Along the first steps of the visual pathway, from the retina to the LGN to the primary visual cortex (Brodmann area 17), a topographic mapping is preserved where nearby information from the visual field is processed in nearby neurons. As the information comes from the retina, we call this a retinotopically correct mapping. Many models for mapping have been proposed, so we will look only at some basic components of such models (for a more comprehensive overview of developmental models, see chapter 10 of Sterratt et al., 2011).

Mechanisms that have been proposed include (a) neural activity, where the firing pattern of neurons includes information about their target location; (b) chemoaffinity, where target cells exhibit chemical labels that, when matching the desired label of the axon, lead to connection establishment; (c) competition, where axon guidance is influenced by interactions with other axons; (d) timing, where both axons and target cells emerge over a longer time window and incoming axons establish connections only with newly formed target-layer cells; and (e) fiber ordering, where axons reach their target depending on their position within the growing fiber tract.

Chemoaffinity is a major component of current models. A biological source of a chemical label could be the Eph receptor with its ligand ephrin as gene knockout experiments resulted in a loss of topographic mapping. Ephrin exists in two forms, A and B. Given two sets of molecules, two axes of a map, such as the horizontal and vertical axes of the visual field, could be encoded (see figure 5.5). Axons with a high concentration of EphA receptors would then connect to neurons in the target layer with a low concentration of ephrinA.

Another component of models is lateral inhibition through neural activity. As these models start from a more random widespread connectivity between regions and then refine connections to form a topographic map, they are also called self-organizing maps. An early model (Willshaw and von der Malsburg, 1976) proposed a topographic map in the retina, as the source, and connections both toward and within the target map. Lateral connectivity in the target map includes both cooperation (positive interaction) and competition (negative interaction) as modeled through a Mexican hat function. Neural dynamics are simulated while each neuron in the source layer is sequentially activated. For each activated source neuron, the most active neuron in the target layer is determined (the winner) and the connection to the winner is

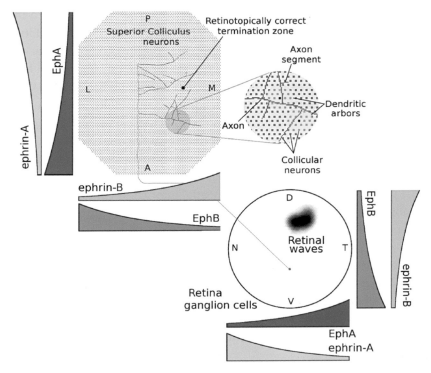

Figure 5.5
Cartoon showing simulated retina and colliculus with morphogenetic fields. Retinal ganglion cells (RGC) extend into the colliculus and arborize there. Axon growth was mediated by molecular gradients and trophic feedback to synapses. Retinal waves drove patterns of RGC spiking. Each axon was composed of a series of axon segments (an example is shown in red in the expanded view), and each segment was able to produce synapses with overlapping dendrites (yellow). Directions: L = lateral, M = medial, A = anterior, P = posterior, D = dorsal, V = ventral, N = nasal, T = temporal (adapted from Godfrey et al., 2009).

strengthened. As a result, a topographic map in the target layer can emerge. Many variations of self-organizing maps have been proposed, including models without lateral connectivity or neural dynamics (Kohonen, 1982) and models with continuous input space using neural field models (Amari, 1977; Qubbaj and Jirsa, 2007; Robinson, 2013).

Indeed, a biologically more detailed computational model indicates that lateral connectivity in the target region, and resulting local competition through lateral inhibition, might not be necessary to form topographic maps (Godfrey et al., 2009). A model of retinotopic development, including axon and synapse growth, molecular guidance, and synapse plasticity, shows that synaptic plasticity is realized by variation in the number of synapses between neurons, not through alteration of individual synaptic weights. In this model, retinal axons enter the anterior side of the colliculus and extend in a largely linear manner to the posterior side (see figure 5.5). Interstitial branches then sprout and extend toward the retinotopically correct area of the colliculus for the given axon, based on chemoaffinity compatibility between each axon and the expression of molecular markers in the colliculus. Activity-dependent trophic feedback mediates growth and retraction of individual synapses, with a trophic factor stabilizing synapses that contribute to spiking activity in the postsynaptic neurons and synapses that receive insufficient trophic feedback retracting. Correlated retinal activity, in the form of retinal waves, provides spatial information allowing synapses from retinal ganglion cells (RGCs) originating from adjacent points on the retina to establish strong connections to collicular neurons which are adjacent to each other as well. Trophic factors enhance axon and synapse growth in the areas of the axon where they are received, and spike-time-dependent plasticity modulates the excitatory strength of individual synapses. For topographic map formation, spike-time-dependent plasticity, gradient detection by axonal growth cones, and lateral connectivity among collicular neurons were not necessary. Finally, instructive cues for axonal growth appeared to be mediated first by molecular guidance and then by neural activity.

Homeostasis
Related to neural activity is the concept of homeostasis. While the environment changes in terms of growth of other connected parts of the brain or the change in sensory input, for example, after birth, one idea is that the activity levels of individual neurons should be kept constant (*homeostasis*). Similar response patterns of neurons could be maintained through changes in ion channel expression and distribution and the establishment or removal of synaptic connections.

6 Layer Formation

6.1 Overview

The gray matter of the brain, the cortex, shows an architecture with multiple layers. Layers were initially determined based on cytoarchitectural features that can be observed through microscopy such as cell type and cell density (Meynert, 1867). However, novel methods can include information about neuronal connectivity, gene expression, and neurotransmitter receptor distributions.

Within the cortex, the number of layers can vary. Most regions are part of the isocortex with six defined layers, numbered from the top of the cortex (layer I) to the boundary between gray matter and white matter (layer VI): layer I, the molecular layer, consists of only a few nerve cells; layer II, the external granular layer, consists of numerous small, densely packed neurons; layer III, the pyramidal layer or external pyramidal layer, is composed of medium-sized pyramidal nerve cells; layer IV, the inner granular layer, contains small, irregularly shaped nerve cells; layer V, the ganglionic or inner pyramidal layer, includes large pyramidal cells; and layer VI, the multiform layer, contains small polymorphic and fusiform nerve cells. While regions show six layers, their organization changes between regions with some layers almost absent or greatly reduced in thickness and other layers showing a more complex organization with different bands of cells within the same layer. For example, motor cortex, with large projection neurons toward the spinal cord, shows an extended layer V while the primary visual cortex (Brodmann area 17) shows an enlarged input layer IV with a subdivision in distinct sublamina (IVA, IVB, IVCα, IVCβ).

There are also regions in the cortex where fewer than six layers occur. The allocortex only shows a single neuron layer and can be further sectioned into the archi- and paleocortex. The archicortex includes the hippocampus, and the paleocortex, also called "rhinencephalon," includes the secondary olfactory center (the olfactory bulb

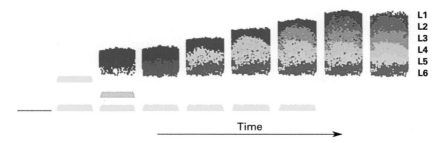

Figure 6.1
Simulation of the sequence of lamination steps. In this simulation of layer formation in the peristriate area of the human temporal cortex, initially only a small set of progenitor cells exists in the ventricular zone (left, dark blue). This proliferative zone generates neuronal precursors in an exponential manner. These precursors, via symmetrical and asymmetrical division, give rise to different cortical layers. The first cells that migrate are marginal zone (MZ) cells (dark blue, top). Subsequently, layer 6 cells (blue) physically push the MZ, constituting a mechanistic process that is repeated for layers 5 (cyan), 4 (green), 3 (orange), and 2 (red). (Adapted from Bauer et al., 2020.)

in the nose is the primary center). Surrounding the allocortex is the periallocortex that already shows multiple neuron layers and consists of the presubiculum, parasubiculum, and entorhinal cortex. Sometimes, one might also distinguish another type of cortex, the proisocortex which comprises the paralimbic cortex, which shares many but not all features of the isocortex.

A better understanding of how layers form during cortical development is relevant from a biological as well as a medical point of view. Many brain pathologies, such as autism (Crawley, 2012), schizophrenia (Nawa et al., 2000), fetal alcohol spectrum disorder (Riley et al., 2011), or epilepsy (Bozzi et al., 2012), have been associated with pathological layer formation during development.

6.2 Cortical Layers

Layer Formation
We can first look at the time course of layer formation. During cortical development, progenitor cells proliferate and differentiate into different neuron types in a consistent sequence. Marginal zone (MZ) cells are the earliest cells to be produced. Afterward, differentiation into layer 6, layer 5, layer 4, and layer 2/3 cells occurs (see figure 6.1).

Cell proliferation and migration is guided through gene regulatory networks. A hypothetical model of these circuits has been described by Bauer et al. (2020). From an initial, purely proliferative stage S1 cells transition to stage S2, leading to neuronal

Layer Formation

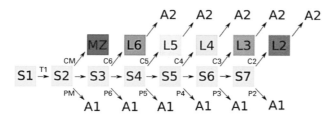

Figure 6.2
Schematic of potential sequence of cortical layer formation. During cortical development, progenitor cells proliferate and differentiate into different neuron types in a consistent sequence. Marginal zone (MZ) cells (dark blue) are the earliest cells to be produced. Afterward, differentiation into layer 6 (blue), layer 5 (cyan), layer 4 (green), and layer 2/3 (purple) cells occurs. Here, a potential gene regulatory network of cortical lamination is shown. Arrows indicate (probabilistic) paths of the gene regulatory network. Cx and Px are probabilities for cell migration and cell death (apoptosis), respectively. At the end of the lamination process, a final apoptotic step (A2) occurs that depends on the local neighborhood of cells. Yellow color indicates stem cell states, that is, these cells have the potential to divide. Dark blue, light blue, cyan, green, orange, and red indicate final cell types (after Bauer et al., 2020).

migration of MZ cells. Alternatively, apoptosis can occur, cells can continue to divide, or the following stage can be reached. Apoptosis can also occur at a later stage after migration when neurons have reached their target cortical layer (for a potential gene regulatory circuit, see figure 6.2).

Apoptosis

Apoptosis can occur at different stages of brain development, caused either by an internal event (programmed cell death) or by external factors (e.g., neurotrophic death). The vast majority of cells undergoing programmed cell death in vertebrates are neurons. However, cell death is not limited to brain development in higher organisms. Even for the nematode *C. elegans*, apoptosis occurs throughout development. For example, 8 out of 33 neurons of the cell lineage ABalpa, mainly responsible for neurons of the ring ganglia and the pharynx, die during early development (Reichert, 1990). Overall, while 302 neurons of the hermaphrodite nematode survive, 105 die during development (Putcha and Johnson, 2004). This apoptosis of 25% of all neurons occurs early on, and neurons often die within an hour of being born (Riddle et al., 1997).

Cell death can control the cell number but also the organization of the nervous system (Kuan et al., 2000). Mutant mice with reduced apoptosis show overrepresentation of distinct neuronal subpopulations but also major malformations of the brain. For lack of different apoptosis factors, neurulation defects prior to the closure of the hindbrain neural tube were observed (Kuan et al., 2000). Therefore, programmed cell death is a crucial component of brain development.

Connectivity within and between Layers
Layers show a distinct cell type and network organization that has suggested functional roles, leading to the idea of canonical microcircuits for information processing (Douglas et al., 1989). Across the neocortex, the same basic laminar and tangential organization of the excitatory neurons has been found along with characteristic morphology and patterns of connections for inhibitory neurons (Douglas and Martin, 2004).

Despite common characteristics, layer architectures vary in different parts of the brain and across species. The structural model distinguishes different types of cortex: the phylogenetically ancient limbic cortices are agranular if they lack layer IV, or dysgranular if they have a nascent granular layer IV, whereas eulaminate cortices have six layers. Observing the type of cortex and cortical development, there is a preferential vulnerability of limbic areas to neurodegenerative and psychiatric diseases (Garcia-Cabezas et al., 2019). Unlike other regions, limbic areas show lower neural density, late myelination despite early completion of cell migration, and prolonged plasticity. Furthermore, limbic areas in the frontal and temporal lobes are more vulnerable to epileptiform activity, and limbic areas are also affected in Alzheimer's and Parkinson's disease early on.

There is also a distinct organization of how layers in one region project to layers in another region. The connections between brain regions, including their existence and laminar origins and terminations, are linked to fundamental structural parameters of cortical areas, such as their distance, similarity in cytoarchitecture, defined by lamination or neuronal density, and other macroscopic and microscopic structural features (Hilgetag et al., 2019). However, cytoarchitectonic similarity between regions is a stronger predictor of connectivity than distance, similarity of cortical thickness, or cellular morphology. Changes in cytoarchitecture—for example, in overall neuronal density—follow spatial gradients and relate to variations in cellular features of regions.

Layers—A Structure without a Function?
Given different types of cells and distinct connectivity patterns, it is tempting to think about layers as functional units—a microscale equivalent of modules that we observe at the macroscale of connectivity between brain regions. However, there is evidence that the neuron types and connectivity but not the spatial organization into layers is crucial for circuit function. The *reeler* mouse mutant lacks the protein Reelin which guides neuron migration and the formation of layers, resulting in a cortex without a distinct

layer architecture. In such a cortex, different neuron types occur across a wide range of cortical depths; however, the exact pattern still varies between different cortical regions (Guy and Staiger, 2017). The total number of neurons, the proportion of excitatory and inhibitory neurons, the cell number for different neuron types, neural connectivity, and neuron morphology remain similar to the wild type with intact Reelin gene. As a result, even without a laminar structure, function and behavior remain largely intact. It therefore seems that layers, as a spatial organization of cells, themselves are not crucial for information processing in cortical circuits while the connectivity and distribution of neuron types are the main factor for cognition (Guy and Staiger, 2017). Note, however, that disruption of the Reelin gene in humans has more severe effects, resulting in a lissencephalic brain with clinical symptoms of ataxia, cognitive delay, and epilepsy. It is difficult to tell whether these functional deficits for humans are due to the changed cortical layer organization or due to the large overall change of having an unfolded cortex.

Layer-Homologue Structures and Evolution
While we focused on cortical layers in mammals, there are related structures in other species (Briscoe and Ragsdale, 2018). In birds and reptiles, the pallium (dorsal telencephalon) forms a large structure called the dorsal ventricular ridge of neuronal cell body cluster (nuclei). For nonavian reptiles, a three-layered cortex lies above this structure. For birds, this cortex is modified into a second nuclear complex known as the Wulst. Despite this wide range of structures, the homologous organization becomes clearer when one is looking at neuronal cell types: while the avian Wulst does not contain layers, the circuitry of excitatory neurons is similar to that of the mammalian neocortex. Indeed, as the previous section on the Reeler mouse showed, a layered architecture is not essential to perform a similar circuit function. Studies of neocortical layer markers have shown expression in many amniotes, with and without the presence of a neocortex, indicating cell type homology across species.

There seem to be limits for layer formation across species. While the surface area size of the neocortex increases 1,000-fold between mouse and human, the cortical thickness only increases by a factor of 2–3, depending on the observed brain region (Rockel et al., 1980).

6.3 Mechanisms for Layer Formation

Simulating Cell Growth
A computational model of neocortical neuronal cell acquisition with parameters for cell cycle length, commitment to cell cycle exit, and cell death could reproduce the

number of neocortical neurons in adult and developing rhesus monkey and human (Gohlke et al., 2007). In this model, there are two populations of neurons, progenitor cells that are still able to divide and postmitotic cells that stop dividing. Both populations can also undergo cell death. This model indicates that the vast increase in neocortex size from rodents to humans can result from an increased period of neurogenesis and lengthening of the cellular processes of cell cycle progression and death. It also predicts that cell death should play a larger role in the primate neocortex. However, the number of initial progenitor cells, the founder population, does not necessarily scale with the total number of cortical neurons that are generated. Assuming similar cell cycle lengths in primate progenitors, the increase in cortical neuronal numbers does not reflect a larger size of founder population (Picco et al., 2018). Furthermore, the timing of a switch in favor of symmetrical neurogenic divisions produces the highest variation in cortical neuronal numbers.

Another feature when moving from rodents to primates, in addition to an increase in neocortex size and cortical thickness, is a relative increase in the size of the upper cortical layers (II–IV). A model proposed by Cahalane et al. (2014) can reproduce the proportion of lower and upper layers starting again with two populations of cells, precursor cells P and differentiated neurons N, whose population size changes over time

$$\frac{dP(t)}{dt} = P(t)\frac{\ln(2)}{c(t)}[1 - 2d(t) - 2q(t)(1 - d(t))]$$
$$\frac{dN(t)}{dt} = P(t)\frac{\ln(2)}{c(t)}[2q(t)(1 - d(t))]$$

where $c(t)$ is the cell-cycle duration, $d(t)$ is the independent probability that each daughter cell dies after a cell division, and $q(t)$ is the probability that a daughter cell is a differentiated neuron which quits the precursor pool and does not undergo further rounds of cell division. Using spatiotemporal anterior–posterior gradients for some of these parameters, the model predicts (1) a greater complement of neurons per cortical column in the later-developing, posterior regions of intermediate and large cortices, (2) that the extent of variation across a cortex increases with cortex size, reaching fivefold or greater in primates, and (3) that when the number of neurons per cortical column increases, whether across species or within a given cortex, it is the later-developing superficial layers of the cortex which accommodate those additional neurons (see figure 6.3).

Since that framework is formulated on an abstract, analytical level, it does not take into account mechanical interactions between cells, or the setting within the spatial environment. Hence, it is problematic to test more direct, mechanistic hypotheses in

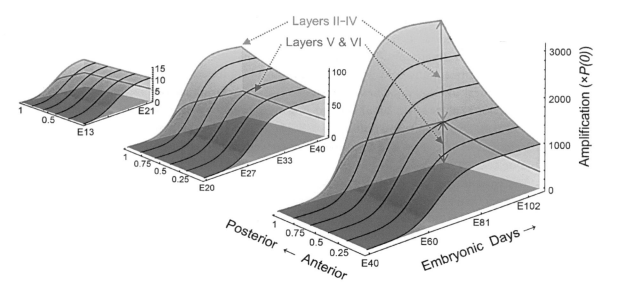

Figure 6.3
Model-predicted interspecies and intracortex differences in the timing, extent, and layer assignment of cortical neuron formation. Shown here are the predicted amounts of neuronal output (in terms of amplification of a unit precursor pool) across the anterior–posterior (spatial) axis of the cortex over the course of embryonic neurogenesis (time axis) for three different cortex scores indicating the duration of brain development time (1.0, similar to a rat; 1.75, similar to a ferret; 2.5, similar to a macaque monkey). The larger cortices have a longer developmental interval, produce orders of magnitude more neurons in total, and, in particular, have a greater complement of upper layer neurons. The anterior–posterior gradient in neuron number becomes more pronounced in larger cortices, and it is the upper layers which accommodate the greater proportion of the increasing quantities of neurons. (After Cahalane et al., 2014.)

such an approach—for example, on the involvement of certain pathways or extracellular cues. A different computational approach is taken by Caffrey et al. (2014), who employ an agent-based model to model cortical layer formation. Newly generated neurons can be either in an initial active state where they can voluntarily change position or in a later passive state where their position can only be changed by mechanical interactions with other cells. Over time, all cells convert to the passive state, and the development simulation ends. Five batches of active cells are generated, forming five cortical layers (layer I is always absent). This model is able to reproduce layer formation for normal and abnormal development. Importantly, the model can reproduce the scrambled cell positioning of the layer-free Reeler mutant mouse.

However, this model does not contain the developmentally crucial aspect of proliferation and relies on highly simplified intercellular interactions. In contrast, Zubler et al. (2013) present an agent-based computational model for cortical lamination with mechanical interactions between cells. Based on this approach (Zubler and Douglas, 2009), cortical development has been studied from a conceptual perspective of self-construction (Zubler et al., 2013) and put in correspondence with genetic data (Zubler et al., 2013). Their simulation framework CX3D includes physical interactions between cells, the growth of axons and dendrites during development, and also gene regulation guiding cell division.

The complex regulatory interactions within cells can be modeled with the G-code framework, which is based on a set of 11 primitives, each representing an elementary cell function, including *move* for the change in soma or growth cone position and *secrete* and *detect* for the production and detection of signaling molecules. During the life of a cell, intracellular agents for different functions can be generated or removed. Using complex models, the generation of cortical layers or the generation of connections between cells can be simulated (see figure 6.4).

Modeling of regulatory circuits can help to explain some counterintuitive developmental phenomena. Brain cancer (glioma) is thought to be due to mutations throughout the lifetime. While one would expect that the older a subject gets, the higher the probability to be affected, there is a dip in the incidence rate after the age of 80 years. A computer model using empirically determined estimates of neural stem cell number, cell division rate, mutation rate, and oncogenic potential could yield results which matched actual demographic data in the human population. In such a model, the decreasing incidence rate at old age was due to an exponential decrease in the number of neural stem cells (Bauer, Kaiser, and Stoll, 2014).

Modeling Layer Formation across Brain Regions and across Species
Using the gene regulatory network structure shown in figure 6.2, the roles of cell division, cell migration, and cell death can be studied systematically. Following an initial exponential proliferation phase, a sequential differentiation phase sets in. Such a model is able to reproduce the variation in the thickness of individual layers within and between species that is seen in anatomical studies (Herculano-Houzel et al., 2008; Collins, 2011). The model also reproduced the sequential development of the layers in the human temporal cortex in the correct order (Zubler et al., 2013).

The growth model can be adjusted to also reproduce the cell numbers in different cortical layers across brain regions and across species (Bauer et al., 2020). These simulations suggest a crucial role for the temporal architecture of apoptosis: in an

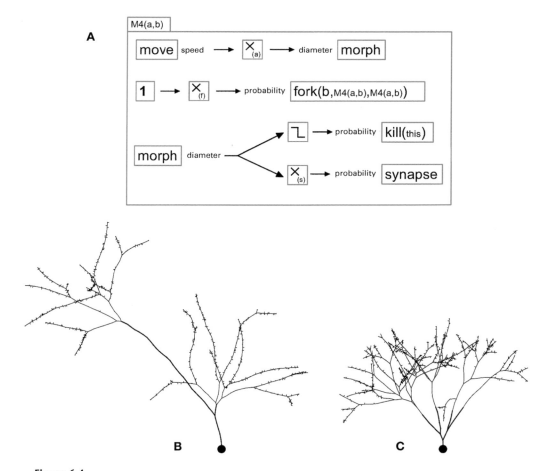

Figure 6.4
Using G-code to specify development simulations. A G-machine capable of producing various branching patterns, with synapse formation. (A) The machine M4 contains four independent processes. (1) A move primitive elongates the neurite and transmits the current speed to a filter where it is multiplied by the coefficient a and fed into a morph for diameter reduction (proportional to the distance traveled). (2) With a probability f, a fork element triggers a neurite bifurcation, producing two daughter branches with diameters smaller than the mother branch's diameter by a factor b, and instantiates in each of them a copy of the machine M4. (3) When the diameter falls below a certain limit, a kill primitive removes the machine. (4) A synapse primitive produces pre- or postsynaptic processes with a probability inversely proportional (proportionality constant s) to the branch diameter. (B) When the diameter is decreased at the bifurcation points but not during elongation, the path from every branch tip to the soma passes the same number of branch points, regardless of the actual length. Note the spines are produced with a density inversely proportional to the neurite diameter. (C) When the diameter is decreased only during elongation and not at branch points, each branch tip is at the same path length from the soma. (After Zubler et al., 2011.)

early apoptotic stage (A1), cell death occurs independently from the local extracellular environment and helps generate roughly the final cell numbers. After layer formation has occurred, a second stage of apoptosis (A2) serves a different role by improving the segregation between layers, hence reducing the overlap among layers. Here, the signals from neighboring cells that convey information on the suitability of cell positions within the extracellular context are crucial. In agreement with this computational model, two distinct types of apoptosis were observed experimentally (Rakic and Zecevic, 2000): embryonic apoptosis, which was synchronous with proliferation and migration of neuronal cells, and later fetal apoptosis, which coincided with differentiation and synaptogenesis.

Altogether, modeling the emergence of layers, and their connectivity within and between regions, involves interactions between cells and genetic circuits guiding cell division, cell migration, and cell death. Recent modeling and experimental results point to a crucial role of apoptosis for ensuring the variety of layer organizations that are observed across regions and species. For experimental studies, layer formation can also be observed in cell cultures even though it is debatable to what extent such approaches represent in vivo development. Future experimental and computational studies, evaluating the role of different genetic and environmental factors, will help to elucidate how layers form and why their architecture can change for developmental brain disorders.

7 Axonal Growth

7.1 Overview

After the formation of layers, neurons start to produce several extensions (neurites), some of which will develop into the dendrite while one will become the axon of the neuron. Determining which neurite develops into an axon follows a competitive process (Goslin and Banker, 1989; van Ooyen, 2011). Growth cones of the axon and the dendrites lead to elongation and branching of the neurites. This results in characteristic morphologies for the shape of different neuron types, from Purkinje cells in the cerebellum with a flat (planar) dendritic tree to basket cells with a dendrite all around the soma of the neuron. Using cell reconstruction techniques, we can observe the shape of many types of neurons in many different species (Ascoli et al., 2007), providing a comprehensive database of single neuron morphology (see figure 7.1).

The direction of axonal growth can be influenced by chemical gradients in the environment with axonal growth cones moving into the direction of a higher concentration. While some chemical substances in the surrounding medium can be attractors of axons, others can be repulsors, directing axonal growth away from the source of the substance. Axons can also follow along the path of existing fibers, a process called fasciculation (Hentschel and van Ooyen, 1999). When axons have arrived in their target region, they can branch, forming axon collaterals and connecting to many different neurons as well as forming multiple synaptic connections with the same neuron.

Further fine-tuning of synaptic connectivity occurs due to competition for target-derived neurotrophic factors where synapses that do not receive sufficient neurotrophic factors from the postsynaptic neuron will be removed and axonal branches will retract. While removal or addition of synapses is called _structural plasticity_ (Butz, Wörgötter, and van Ooyen, 2009), the adjustment of the weight of an existing synapse is called _functional plasticity_. Both establishment and refinement of connectivity are influenced by neuronal activity (van Ooyen, 2011).

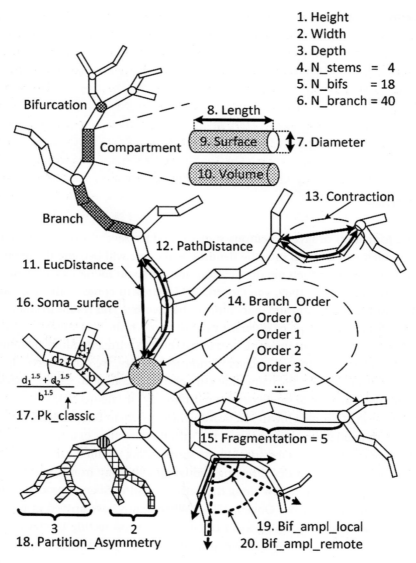

Figure 7.1
Measurements using the L-Neuron tool (Scorcioni et al., 2008) based on reconstructed morphology of neurons within the NeuroMorpho database (after Costa et al., 2010).

Axonal Growth

Another process related to neural activity is the development of a myelin sheath around axons, provided by oligodendrocytes within the central nervous system. Myelination of axons is an ongoing process starting around the time of birth in humans and continuing right into adulthood. Even though there is some ongoing activity before birth, such as spontaneous retinal waves, the massive increase of external sensory input and the growth of brain size after birth drive the increase in myelination. Increased myelination of axons speeds up information propagation by a factor of 10: whereas action potentials travel with a speed of 0.1 m/s for unmyelinated axons, speed increases to 1 m/s for myelinated axons of the same axon diameter (excluding the myelin sheath that is wrapped around the axon). Another option to increase speed without myelination, such as the increase in axon diameter—for example, in a squid giant axon—is prohibited in mammals by the volume within the skull that is available for the brain (see also chapter 10 on cortical folding).

7.2 Establishing Links between Neurons

How Is Wiring between Neurons Encoded?

There are several possible explanations for how an exponential distance dependence may arise (see chapter 11 in van Ooyen, 2003). For example, the target of a connection could be genetically encoded by axon guidance molecules (Borisyuk et al., 2008). Under such a scenario, the target region secretes guidance molecules that diffuse through the tissue. Since there is a higher concentration near the source where the molecule is secreted than at distant regions, a concentration gradient arises, and axons can travel in the direction of higher concentration, toward the target region. The concentration of molecules deposited at one location decays exponentially with time and with distance away from that location (Murray, 2003). This relationship is given by

$$c(x,t) = \frac{Q}{2\sqrt{\pi Dt}} e^{\frac{-x^2}{4Dt}},$$

where D is the diffusion coefficient and Q is the initial amount of particles per area. Given a threshold for the detection of guidance molecules, neurons closer to the source will be more likely to pick up the signal than neurons further away, leading to distance dependence in the establishment of connections. However, this model would incur a hard border—a distance beyond which the concentration falls below the threshold and where no connections occur, whereas for distances shorter than the critical distance, the probability of establishing a connection would be close to 100%. Under this regime, an inverse sigmoid distribution of connection lengths would be expected, which is unlike the curves that are found in neural systems. In addition, guidance through attraction and repulsion is usually discussed for the global level of long-distance connectivity but not for the local connectivity within regions. Whereas guidance cues are a defining factor at the global level, their role on the local level is limited because of the problem of establishing a stable gradient for very short distances (less than 0.7 mm). According to models (Goodhill and Urbach, 1999), the ability of a growth cone to follow a gradient of a ligand molecule depends on the concentration of the ligand, the speed of ligand diffusion, the size of the growth cone, and the time over which it averages the gradient signal with the minimum detectable gradient steepness for growth cones in the range of a 1% to 10% concentration difference across the width of the growth cone.

The standard hypothesis for guidance cues is that they are provided by controlled gene expression, specific transmitter-receptor systems, and growth factors (Sperry, 1963; Yamamoto et al., 2002). This hypothesis involves diffusion-based mechanisms. Recent studies have investigated the genetic component of connection development in *C. elegans* (Kaufman et al., 2006; Baruch et al., 2008) and found that a substantial amount (ca. 40% on average) of the variability of connection patterns in *C. elegans* cannot be accounted for by differences in gene expression patterns. An alternative hypothesis, called "Peters' Principle" by Braitenberg and Schüz (1998), suggests that neural outgrowth is basically random, and that specificity in the wiring is derived from the overlap of specific neuronal populations (Binzegger et al., 2004; van Ooyen et al., 2014).

It is important to remember how powerful the genetic influence for fiber tract formation can be. In an experiment where mice were lacking the guidance molecule Semaphorin-6A, axons from the visual part of the thalamus are misrouted and fail to reach the visual cortex, which is instead innervated by somatosensory thalamic axons. Nonetheless, many visual thalamic axons find alternative routes to finally reach the visual cortex a few days later, even leading to the removal of initial somatosensory thalamic axons (Little et al., 2009). This indicates a highly specific targeting mechanism for forming connections irrespective of early initial thalamic axonal pathways.

Finally, another factor for influencing axon growth is mechanosensing. Embryonic brains show stiffness gradients, and, for example, retinal ganglion cells grow toward softer tissue (Koser et al., 2016). The same study also showed that preventing mechanosensation of the growing axon results in aberrant axonal growth and pathfinding errors. Therefore, both mechanical and chemical cues seem to influence axonal growth.

Nonuniform Wiring

As each cortical layer contains a different distribution of neuron types, born at different stages during development, connectivity differs between these structures. However, even when one looks at the lateral connectivity between different patches of the cortical surface, connectivity is not uniform for all species. Injecting neural tracers into one section of the cortical surface leads to labeling of axons and soma of cells in distinct patches away from the injection site. This "patch system" or "daisy architecture" (Douglas and Martin, 2004) may extend over several square millimeters (Rockland and Lund, 1982; Gilbert and Wiesel, 1989; Lund et al., 2003). Patches are most prominent in the superficial layers 2 and 3 where pyramidal neurons establish lateral connections that support clusters of boutons, suggesting that patches are linked to clusters of neurons (Bauer, Zubler, et al., 2014). Across species, from cat and ferret to macaque and human, as well as across brain areas (Muir et al., 2011, table 1), the scaling of patch diameter to distance between patches is remarkably regular, suggesting that patches are an organizational principle of local cortical connectivity (Douglas and Martin, 2004). Furthermore, based on optical imaging, patches seem to correspond to functional domains in terms of neuronal activity (Muir et al., 2011). Patches form a hexagonal lattice that is more homogeneous and regular within the visual cortex but less so in other parts of the cortex (Muir et al., 2011).

Myelination

Lateral connectivity within cortical layers only extends over a couple of millimeters. Connectivity between brain regions, however, can lead to fiber tracts that, for humans and nonhuman primates, are several centimeters long. For long-distance connectivity, myelination is crucial as an unmyelinated fiber tract that is 10 cm long would result in a transmission delay of 100 ms, compared to a delay of 10 ms for a myelinated fiber. Indeed, there is evidence that longer fiber tracts show a higher axon diameter, leading to faster conduction speed in order to reduce transmission delays (Innocenti et al., 2014). However, other factors such as the activation threshold of the presynaptic neuron or the type of postsynaptic neuron of a fiber might also influence the resulting delay in information propagation from one area to another. Furthermore, there are shorter conduction delays from motor, premotor, and somatosensory cortex than from the visual and most of the association cortex (Tomasi et al., 2012). As a result, there is a wide range of delays, potentially indicating a hierarchy of processing speeds. There is increasing evidence that myelination is finely tuned to ensure the correct timing of signal propagation (Salami et al., 2003). Finally, in addition to oligodendrocytes that modulate myelination, astrocytes attached to the node of Ranvier of myelinated axons might modulate conduction speed (Fields et al., 2015).

The myelination of axons, in both the central and the peripheral nervous system, is related to the activity of the neurons, whose axons are being wrapped with a myelin sheath, and to axon diameter (Tomassy et al., 2016). Within the peripheral nervous system, axon caliber correlates with the thickness of the surrounding myelin sheath, and only axons that are thicker than 1 µm become myelinated. In the central nervous system, already axons with 0.2 µm become myelinated. However, myelination is not uniform across the axon with axons from neurons in superficial cortical layers displaying myelinated segments that are interspersed with long, unmyelinated tracts (Tomassy et al., 2014). Such changes along fiber tracts might also be related to the changes in fractional anisotropy along tracts observed with diffusion imaging leading to the approach of connectometry for studying changes in connectivity between regions (Yeh et al., 2016).

Concerning the role of neural activity, optogenetic stimulation of projection neurons of the mouse motor cortex increases proliferation of oligodendrocyte progenitor cells and differentiation of oligodendrocytes in both cortex and subcortical white matter. As a result, the thickness of the myelin sheath is increased along the whole length of axons of stimulated neurons, not just near the stimulation site (Gibson et al., 2014).

Such an interplay between activation and myelination also seems to occur in humans in both adults and children. Diffusion imaging studies showed changed fractional anisotropy of language-related fiber tracts in bilingual compared to monolingual children (Mohades et al., 2012). Similar changes were observed for adults who learned Chinese as a second language (Schlegel et al., 2012; Hosoda et al., 2013), suggesting that activity-related changes in myelination can occur throughout the life span right into adulthood. Fiber tract changes were also observed after learning new motor skills such as piano playing or juggling (Bengtsson et al., 2005; Scholz et al., 2009).

There are three different cellular sources for increased myelination in the central nervous system (Long and Corfas, 2014): (1) an increase in myelin production of already existing oligodendrocytes, (2) the transformation of oligodendrocyte progenitor cells, which are located both within gray and white matter and exist both in child- and adulthood, into oligodendrocytes, and (3) adult neural stem cells—for example, in the dentate gyrus and subventricular zone—that can give rise to oligodendrocytes.

7.3 Mechanisms for Axonal Growth

Generative Models versus Simulated Models

One approach to understand the role of different factors for connectome development is to use generative models. These models use a single factor or combinations of factors to generate connectivity that is close to the observed connectivity in macaque and *C. elegans* (Costa, Kaiser, et al., 2007), mouse (Henriksen et al., 2016), or human (Betzel et al., 2016). However, one should keep in mind that these studies are based on features of the adult network. The contribution of factors might be different during development as brain sizes are smaller, not all brain regions already exist, and the relative spatial position of regions or neurons prior to cortical folding or migration differs. Concerning topological factors, not all connections already exist and some early connections might disappear later during development as was found for transient projection fibers (Rakic and Riley, 1983; Innocenti and Price, 2005; Luo and O'Leary, 2005).

Alternatively, one might use computer simulations for observing brain network growth where the number of nodes is increasing over time and where the spatial position of nodes and links may change over time. Finally, models may observe the formation and removal of synaptic connections and the growth of dendritic trees.

Straight versus Curved Growth

Axons tend to grow in a straight line (Borisyuk et al., 2008) while extensions of the axonal growth cone, filopodia, search the space in front of the growth cone for potential neurons with which to connect. The direction of axon growth might change because of physical obstacles (neurons or other tissue) and adhesion (Franze, 2013). It might also change because of chemical factors, molecules that either attract or repulse the growth cone (Sperry, 1963).

On average, axons tend to connect to nearby targets with the probability that two neurons are connected exponentially decreasing with the distance between them. Such an exponential decay of the axon length probability can be observed both at the local level of axonal connectivity between pyramidal cells within rat layers II and III of the primary visual area (Hellwig et al., 1994; Hellwig, 2000) (see figure 7.2A) as well as at the global level of structural and functional connectivity between brain regions (Kaiser and Hilgetag, 2004a; Kaiser et al., 2009; Kaiser, 2011; Ercsey-Ravasz et al., 2013). But why is growing in a straight line the preferred mode of forming axons, and how does this influence how long axons grow before establishing a synapse?

There are (at least) two factors that influence how long it takes an axon to reach another neuron. The first factor is the neuronal density as a higher number of neurons per volume makes it more likely to hit a potential target neuron. Say that for each volume element (e.g., 1 µm^3) there is a probability p that the space contains a neuron and a probability $q = 1 - p$ that the space is empty. Hitting another neuron three volume elements away from the starting point given a (certain) direction means two times passing through empty volume elements and one time (the last growth step) entering a volume element that contains a neuron (see figure 7.2B): the probability to hit another neuron after n steps or passed volume elements is $P(X = n) = q^{n-1} * p$. The exponential

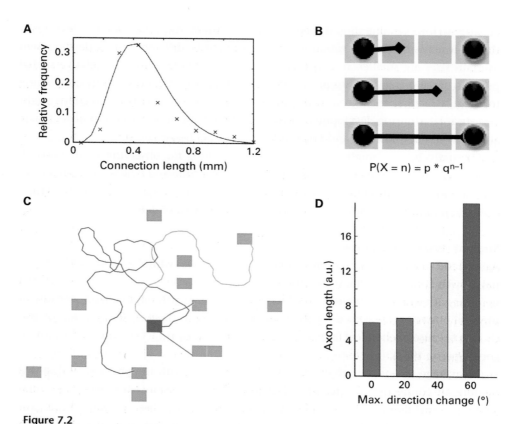

Figure 7.2

Formation of connections. (A) Most axons within layers 2/3 of the rat primary visual cortex connect to nearby neurons, and only few go over a longer distance (redrawn from Kaiser, 2011). (B) This exponential decay with distance can be explained in that the probability to reach another neuron after n steps is the probability to not reach a neuron for $n-1$ steps (q^{n-1}) times the probability to reach a neuron in the final step (p). (C) Often changing direction during axon growth, comparable to a random walk, leads to the growth cone visiting spatial regions that have already been investigated before. (D) This leads, on average, to longer lengths of formed axons (adapted from Kaiser, 2017).

decay of the probability to encounter another neuron as a function of distance between two neurons (or n steps) means that it is more likely to hit a nearby neuron than hit a neuron that is far away after failing to encounter many nearby neurons along the way. Imagine going through a crowded room in a straight line. You are more likely to bump into someone early on than only passing through an empty space and reaching a person who is farther away. Therefore, basic considerations for axon growth can already account for the observed exponential decay of connection probability with distance (Kaiser et al., 2009).

The second factor that influences axon length is the degree of curvature during axon growth. Curvature can range from the extremes of a straight line (without being curved) to a random walk where the axon can arbitrarily change direction at each step (see figure 7.2C). Interestingly, often changing the direction of axon growth does not increase the chances to reach another neuron. For straight growth, each move of the growth cone enters a space that has not been explored before. For a change in direction, as part of the next step, the growth cone is partially moving into space that has already been explored. The resulting curved axon is both using more time and a longer axon length before reaching another cell. An exemplary simulation of axon growth shows that the average length of an axon between two neurons, directly corresponding to the time it takes to establish that connection, is twice as high compared to straight growth if the growth cone is allowed to change direction up to ±40° at each step (see figure 7.2D). Growing axons along a straight trajectory is beneficial in terms of wiring length but also for forming functioning circuits: for example, active turning toward an attractor line is needed to get the correct connectivity in the developing tadpole spinal cord (Li et al., 2007).

While axons normally grow in a straight line, their target neurons might often be off course, leading to the need to change the growth trajectory. Causes for changing direction could be physical obstacles that either block the path or lead to adhesion of the axon. Alternatively, concentration gradients can lead an axon toward a target. In this way, axonal growth cones detect the concentration of a molecule and follow a gradient of increasing concentration toward the source of that molecule (Song and Poo, 2001). However, the concentration of a molecule decays exponentially with the square of the distance between source and growth cone. Given the diffusion constant in vivo (Goodhill, 1997), the estimated maximum distance across which a growth cone can detect molecules emitted from a target neuron is 1 cm (Goodhill and Urbach, 1999).

How can fibers be guided to the correct off-course target if the target is more than a centimeter away? Many fiber tracts between brain regions in humans are longer than 10 cm, not to mention axons through the spinal cord that can be longer than 1 m.

There are two related strategies: axons can follow pathways that were established earlier by pioneer neurons, or neurons can connect at an early stage of development when the total size (scale) of a neural system is small.

Fasciculation

Fasciculation is the mechanism whereby a small number of pioneer neurons form pathways that guide the axons of the following neurons, resulting in a bundle of axon fibers. This might also be the case for the nematode *C. elegans* (White et al., 1986; Durbin, 1987) where some neurons in the ventral cord are formed early on (Varier and Kaiser, 2011), providing a pathway between anterior and posterior parts of the worm. Also, as observed for *Drosophila* (Ito et al., 2013), neurons of the same clonal lineage often project to other regions through bundled fibers along the same trajectory. Computational models of fasciculation have been developed (Hentschel and van Ooyen, 1999; Hentschel and van Ooyen, 2000) and compared with experimental findings. For example, a model of directed random growth along a gradient with attachment between adjacent fibers can reproduce observed fiber tract patterns in the olfactory bulb (Chaudhuri et al., 2011).

The reliance on pioneer fibers might prevent more diverse connectivity to other areas located afar. Fasciculation reduces the amount of guidance that is needed during development as only the pioneer fibers need to be guided to the correct location whereas later fibers attach to the existing tract through adhesion and follow it toward the target region.

Scaling

The alternative to fasciculation, and a potential mechanism for the growth of pioneer fibers, is to form connections early on when the brain size is at a smaller scale and potential target neurons are nearby. Studies in *C. elegans* have shown that 70% of long-distance connections exist between pairs of neurons that are both born early before hatching (Varier and Kaiser, 2011). At this early time, the body size is less than 20% of the adult size, which means that the maximum distance between any two neurons is less than 0.2 mm. This makes axon guidance, as indicated by the increased expression of guidance molecules such as Netrin and Nerfin-1 around this time, feasible.

Based on the maximum gradient distance of 1 cm that axonal growth cones can pick up, scaling (establishing connections while distances are small) will be the preferred mechanism during early connection formation, especially for forming pioneer fibers, while fasciculation (following existing pioneer fibers) will be the main mechanism at later stages of development when source and projection target are further away.

Branching

Axons form branches once they reach their target destination. The technique of Dynamic Time Warping has been developed to observe the time course of how axonal branches can be extended, retracted, or removed (Chalmers et al., 2016). For RGCs, the branch birth rate increases over time as axons get closer to their target (see figure 7.3). Blocking neural activity results in a higher branch death rate but does not affect the branch birth rate (Chalmers et al., 2016).

A recent study of branching patterns of thalamic afferents in cat area 17, fundamental for the synaptic connections of the ocular dominance stripes, has observed different rules for branching that optimize for segment length distribution as well as for a length-weighted asymmetry quantification of the axonal morphologies. This combined optimization can hence capture global morphological properties of the entire data set, as well as morphologies of individual axons. As a result, the model surpasses the statistical accuracy of the Galton-Watson model, the most employed model for biological growth processes (Fard et al., 2019).

Establishing Connections

The axonal growth cone needs to be close to a potential target neuron in order to form a synaptic connection. The position of axons, and their overlap with the position of dendrites, is often sufficient to explain the pattern of synaptic wiring on the dendritic tree (Li et al., 2007). However, even when this condition has been met, the formation of a connection does not automatically follow. The frequency with which nearby axons and dendrites form a synaptic connection, the filling fraction (Stepanyants et al., 2002), ranges from 50% in the tadpole spinal cord (Li et al., 2007) to 12% in macaque visual area V1 (Stepanyants et al., 2002). For example, confocal microscopy and recordings in L5 pyramidal neurons of rat somatosensory cortex have shown that less than one quarter of touches between connected neurons led to the formation of a synapse. In total, only 10% of all possible synapses for adjacent axons and dendrites were established (Kalisman et al., 2005).

One mechanism that can explain why only a fraction of potential synapses can be established is competition for space on the dendritic tree (Kaiser et al., 2009). In this model, a growth cone can only form a synapse with a nearby neuron if the closest position on that neuron, the "docking space," is not already occupied by another synapse. For the occupied condition, on the other hand, forming a connection is impossible. Using a simulation of axon growth, axons would continue growing until either a synapse with another neuron could successfully be established or the embedding space of the simulated tissue was reached. In the simulations that included potential

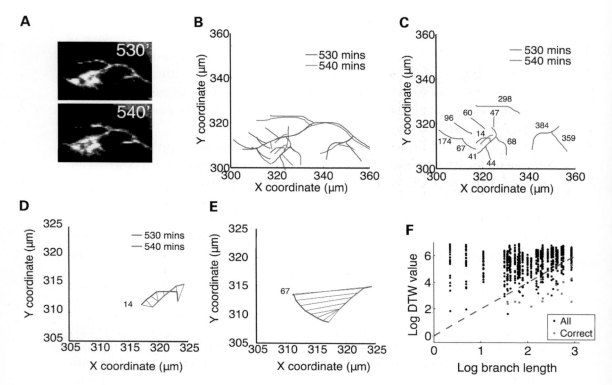

Figure 7.3

Simulating axon branching through Dynamic Time Warping (DTW). (A) Stills from a time-lapse video at 530 and 540 minutes. Brn3C+ RGC axons were genetically sparsely labeled with mGFP. (B) Overlapped axon tracings at 530 minutes (blue) and 540 minutes (red). (C) Comparison of branches from 530 minutes (blue) to one branch from the 540-minute frame (red). For clarity, the primary axon shafts are not shown. The number next to each blue branch is the DTW cost between it and the one red branch. Branches that are likely to be the same between two consecutive traced frames have the smallest DTW distance values. (D) DTW joins coordinates on the two branches (black lines) and assigns the lowest value (14) to the warping path, suggesting a correctly matched branch. This was confirmed by visual assessment of the tracings. (E) DTW comparison between the selected branch from 540 minutes and a branch further away gives a higher DTW value. (F) The log of the DTW distance between pairs of branches near the end of the time-lapse movie (1,700 minutes and 1,710 minutes) versus the log of the length of the branches. The length of several branches could be similar, resulting in columns of matches that nearly overlap. The red points indicate correct matches as determined by visual inspection. The blue dotted line gives the maximum value for an acceptable branch match between consecutive frames, DTW = Length2, (the length of the branch in the first frame). This was included as a threshold for the automated matching of branches and improved the reliability of the DTW algorithm. The branches with no corresponding red dot were correctly identified (compared to visual inspection) as having no match (deleted) (after Chalmers et al., 2016).

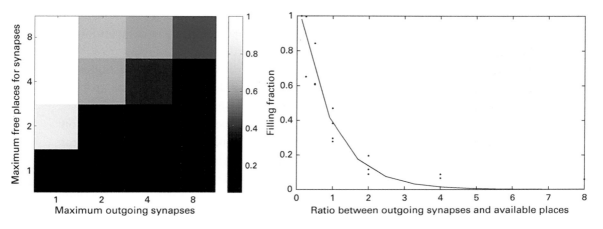

Figure 7.4
Influence of competition on the filling fraction. (A) Variation of axon collaterals (outgoing synapses) and available places for synapse establishment. The median filling fraction of 20 generated networks with 1,000 neurons for each parameter pair is indicated by the gray level. The maximum number of outgoing synapses denotes how many connections a single neuron can maximally establish. The number of maximum free places for synapses shows how many places each neuron has available for incoming connections. (B) The x-axis shows the ratio between outgoing axon collaterals and maximally available places at potential target neurons. The filling fraction decays with this ratio and follows an exponential function $f = 1.125e^{-r}$, $R^2 = 0.91$ (solid line). (After Kaiser et al., 2009.)

occupation, the filling fraction decreased with increasing numbers of neurons. We can also look at scenarios where the number of outgoing synapses of each axon and the number of available target spaces for each neuron differs (see figure 7.4). Having more outgoing synapses than free places led to more competition and therefore to a lower filling fraction (<0.2). More free places than outgoing synapses, on the other hand, reduced competition and led to a higher filling fraction (>0.6). Testing different scenarios showed that the filling fraction f depended on the competition factor r (ratio between outgoing synapses and maximum free places) following an exponential decay $f = 1.125e^{-r}$.

This model can reproduce filling fractions that were experimentally found in anatomical networks and leads to testable predictions for the development of network connectivity (Kaiser et al., 2009). For example, a decrease in axon collaterals or an increase in space on target neurons should reduce competition and therefore lead to higher filling fractions. Conversely, smaller dendritic trees, with the neuron density remaining the same, should lead to reduced filling fractions. Indeed, a relationship has been found between the size of the dendritic tree and the number of innervating axons that survive into adulthood (Purves and Lichtman, 1980), pointing to competition for space in the developmental pruning of axonal connections.

In an alternative model for the filling fraction (Stepanyants et al., 2002), nonestablishment of synapses was explained by axons running close to the dendritic tree of a neuron but missing the dendritic spine in their straight-forward movement. However, whereas this model numerically reproduced the experimentally found filling fraction, the explanation may not be biologically realistic: Filopodia of axonal growth cones can be up to 40 µm in length (van Ooyen, 2003) potentially covering up to 80 µm in the forward direction searching for potential binding places. The distance between spines on the dendritic tree, however, is significantly smaller than this search space, with typically two spines per micrometer for dendritic segments of cortical pyramidal cells (Braitenberg and Schüz, 1998). Competition might therefore provide a more realistic explanation for the experimentally observed filling fractions.

Role of Spatial Borders

We simulated mechanisms of spatial growth, in such a way that connections among nearby nodes (i.e., areas) in the cortical network were more probable than projections to spatially distant nodes (Kaiser and Hilgetag, 2004a). At each step of the algorithm a new area was added to the network until reaching the target number of nodes (55 areas for simulated cat and 73 for simulated macaque cortical networks). New areas were generated at randomly chosen positions of the embedding space. The probability for establishing a connection between a new area u and existing area v was set as

$$P(u,v) = \beta e^{-\alpha d(u,v)},$$

where $d(u, v)$ was the spatial (Euclidean) distance between the node positions, and α and β were scaling coefficients shaping the connection probability. If a new node did not manage to establish connections, it was removed from the network.

The parameter β ("density") served to adjust the general probability of edge formation and was chosen from the interval [0; 1]. The nonnegative coefficient α ("spatial range") exponentially regulated the dependence of edge formation on the distance to existing nodes. The algorithm allowed some nodes to be established distant to the existing network, although with low probability. Subsequent nodes placed near to such "pioneer" nodes would establish connections to them and thereby generate new highly connected regions away from the rest of the network. Through this mechanism multiple spatial clusters were able to arise, resulting in networks in which nodes were clustered topologically as well as spatially.

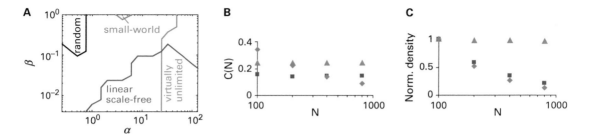

Figure 7.5
Exploration of spatial growth with and without borders. (A) Overview of network types for different spatial growth parameters ($N = 100$ nodes). Low values of α made edge formation independent from distance and resulted in random networks. For large values of α only nodes near the existing network could establish connections, and the hard borders were not reached (virtually unlimited). The area labeled linear scale-free was a region in which networks were sparse, were highly linear (including chains of nodes), and showed a scale-free degree distribution. Only a small part of the parameter space displayed properties of small-world networks. (B, C) Comparison of the dependence of clustering coefficient C(N) and density on network size (number of nodes, N). For the simulated networks the clustering coefficient remained constant for limited (red triangles, $\alpha = 5$, $\beta = 1$) and virtually unlimited (blue boxes, $\alpha = 200$, $\beta = 1$) spatial growth, but decreased for growth with preferential attachment (green diamonds). Density was independent of network size only for limited spatial growth. (After Kaiser and Hilgetag, 2004b.)

In a small interval of intermediate values for α ($\alpha \approx 4$, $\beta = 1$), networks exhibited properties of small-world networks with a clustering coefficient that was 39% higher than in random networks (see figure 7.5A). For higher edge probability ($\beta \to 1$), a noteworthy difference between limited and virtually unlimited growth became apparent (Kaiser and Hilgetag, 2004b). While it was impossible to generate high network density under virtually unlimited growth conditions, the introduction of spatial limits resulted in high density and clustering, as well as low characteristic path length. This was due to the fact that, in the virtually unlimited case, new nodes at the borders of the existing network were surrounded by fewer nodes and therefore formed fewer edges than central nodes within the network. In the limited case, however, the network occupied the whole area of accessible positions. Therefore, new nodes could be established only within a region already dense with nodes and would form many connections.

If we take snapshots at different times during network growth, observing the evolution of the network, we can distinguish different types of network generation (see figure 7.5B, C). For preferential attachment (Barabasi and Albert, 1999), both clustering coefficient and edge density decrease as the network grows bigger. For unlimited spatial growth, the clustering coefficient remains stable while the edge density decreases with network size. Only for limited spatial growth do both clustering coefficient and edge density remain stable across different developmental stages.

Testing whether we can replicate cat and macaque structural connectivity (Young, 1993; Scannell et al., 1999; Hilgetag, Burns, et al., 2000), the biological networks featured even stronger clustering than the ones resulting from simulated spatial growth. However, better matches could be generated by extending the local range of high connection probability, so that $P = 1$ for Euclidean distances of $d_{cat} < 0.18$, $d_{macaque} < 0.11$, and P decaying exponentially as before for larger distances (this was implemented by setting $\alpha_{cat} = 5$, $\alpha_{macaque} = 8$, and for both networks $\beta = 2.5$ and thresholding probabilities larger than one to one). The modified approach therefore combined specific features of the biological networks with the general model of limited spatial growth. This yielded networks with distributed, multiple clusters and average densities of around 30% for simulated cat and 16% for simulated monkey connectivity. Moreover, these networks had clustering coefficients of 50% and 40%, respectively, very similar to the biological brain networks (Hilgetag, Burns, et al., 2000). Comparison of the biological and simulated degree distributions, moreover, showed a significant correlation (Spearman's rank correlation $\rho = 0.77$ for the cat network, $P < 0.003$; and $\rho = 0.9$ for the macaque network, $P < 0.00002$).

In contrast to limited growth, virtually unlimited growth simulations with high β resulted in inhomogeneous networks with dense cores and sparser periphery. It is difficult to imagine realistic examples for strictly unlimited development, as all spatial networks eventually face internal or external constraints that confine growth, be they geographical borders or limits of their energetic and material resources. However, virtually unlimited growth may be a good approximation for the early development of networks before reaching borders.

Self-Organization of Patchy Connectivity

Spatial borders, discussed in the previous section, can explain higher clustering but not necessarily the rise of distinct patches, the "daisy" architecture of lateral connectivity across the cortical sheet. Some have argued that functional mechanisms—for example, the response to visual stimuli in the visual cortex—shape patchy connectivity. However, changes leading to patchy connectivity can already be observed at early stages, before afferents reach superficial layers and can bring information about external stimuli (Bauer, Zubler, et al., 2014).

In a computational model of patch formation, early neuronal precursors in the cortex release morphogens that are transcription factors that interact with the genome of other neurons following Gierer-Meinhardt reaction–diffusion dynamics (Turing, 1952; Gierer, 1988). These precursors establish a two-dimensional periodic pattern of morphogens leading to clusters of neurons with similar profiles inherited from their progenitor cells that migrated to form the superficial cortical layers. When these neurons have reached their target layer, the growth cones of lateral axons will seek distant targets with a similar morphogen expression profile, resulting in a patchy organization (Bauer, Zubler, et al., 2014). This model indicates that patchy connectivity could result from processes that occur early during development, without the need for neuronal activity.

8 Formation of Hubs

8.1 Overview

Many real-world networks contain highly connected nodes called hubs. Such hubs play a crucial role for distributing or collecting information. Given that they are connected to a wide range of brain regions, they are also important for synchronization within networks. Finally, as we will see within this chapter but also within part III, they are involved in many brain diseases.

Hubs can be determined through the number of connections of a node, the node degree, but also through other centrality measures, such as betweenness or strength (see chapter 2). Unfortunately, there is no real definition of when a node is called a hub. For a random network, in which each node has, on average, 10 connections, the most highly connected nodes might have just 12 connections. In other words, nodes with the maximum number of connections within a network may or may not be highly connected. Another possibility would be to define hubs as nodes that are connected to at least 10% of all nodes of a network. This works well for dense networks, but for sparse networks, where each node has only few connections, there might be no hubs under this definition.

An alternative to defining hubs is to look at the degree distribution, with the degree given by the number of a node's connections, to search for nodes that are more connected than the average node in the network. For this, one could use the coefficient of variation, CV:

$$CV = STD/M$$

where STD is the standard deviation for the degrees of all nodes and M is the average value. A very broad distribution would lead to a high value of CV so that there are nodes that are several standard deviations more connected than the average node in that network. A very narrow distribution, on the other hand, would lead to a low CV

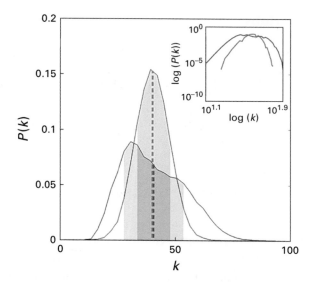

Figure 8.1
Exemplary distributions for node degree k (inset: log–log plot) for networks from linear (red) and nonlinear growth (blue). Linear growth corresponds to the scenario where the network size increases linearly, that is, only one node is added at each time step. Shaded areas show the standard deviation around the mean degree (dashed line). Nonlinear growth yields a wider distribution with more hubs but also more sparsely connected nodes (after Bauer and Kaiser, 2017).

value so that all nodes have a degree that is very close to the average degree of the network (see figure 8.1).

Given this approach, the z-score of a node, the number of standard deviations that its degree is higher than the average node degree of the network, could be a measure of its "hub-ness." Curiously, there is no term in the literature for nodes that are significantly *less* connected than the average node (could we call them "antihubs" or "loners"?). In any case, we will only observe hub formation within this chapter.

8.2 Hub Architecture of Brain Networks

Hubs: Types, Locations, and Roles

Hubs may be classified as provincial (intracluster) or connector (intercluster) hubs (Guimera and Amaral, 2005). Provincial hubs link vertices primarily within a single cluster whereas connector hubs link multiple clusters to one another (see figure 8.2A).

Formation of Hubs

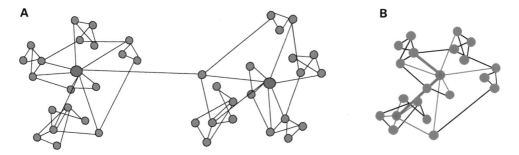

Figure 8.2
Hub features of complex networks. (A) A provincial hub (right) is connected mainly to nodes in its own module, whereas a connector hub (left) is also connected to nodes in other modules (from Bullmore and Sporns, 2009). (B) In a rich-club network, connections between hub nodes (red) are stronger than would be expected (after Kaiser, 2017).

Furthermore, a third type of hub, a bridge hub, would connect multiple modules (Fornito et al., 2015). For the cat and macaque, many hub regions were functionally already known as polysensory or multimodal regions (Sporns et al., 2007).

Deletion of provincial hubs or connector hubs have distinct effects on the small-world organization of the remaining network (Sporns et al., 2007). Deletion of connector hubs disconnects functional clusters, resulting in an increase in the small-world index (Humphries and Gurney, 2008) as shortest paths between disconnected compartments are smaller than in the original connected network. Deleting provincial hubs, on the other hand, disturbs the functional integration within their clusters and leads to a reduction in small worldness due to a reduced clustering coefficient (Sporns et al., 2007).

Macaque area 46 is a prime example of a connector hub. It receives polysensory input from the posterior cortex and also has a large number of connections to distant parietal areas. Another connector hub, the frontal eye field, shares many characteristics with area 46 but is more focused on regions with visual function. Area V4, an example of a provincial hub, integrates information from different parts of the visual system, and lesions in V4 result in deficits that do not rely on a single visual modality (Sporns et al., 2007).

For the cat, cortical hub regions include anterior ectosylvian sulcus, agranular insula, area 7, granular insula, lateral area 5A, and posterior cingulate cortex, based on Kaiser, Martin, et al. (2007), and, concerning subcortical regions, area 35 (perirhinal area), and 36 (ectorhinal area) (Kaiser, Martin, et al., 2007; Sporns et al., 2007).

Functional connectivity hubs in fetal human brains are cerebellum, primary visual and motor regions, as well as association regions (inferior temporal gyrus and medial temporal lobe), all of which show early maturation and myelination during brain development (van den Heuvel et al., 2018).

Rich-Club Networks

Hubs are often more strongly connected than would be expected, forming a rich-club network (Zhou and Mondragon, 2004) (see figure 8.2B). This set of highly interconnected hubs can function as a backbone of the network that allows for the quick transmission and integration of information across clusters. Looking at brain function during tasks, rich-club nodes show a higher outgoing effective connectivity compared to the rest of the network. This connectivity reaches both peripheral nodes and nodes that are active across different tasks. Altogether, this indicates that rich-club nodes support information exchange among peripheral nodes in a task-specific manner while receiving information from the rest of the network (Senden et al., 2018). While a rich-club organization also occurs for technical communication networks, it is absent for some biological systems such as protein-protein interaction networks.

Deviations from normal brain development have distinct effects on the brain's rich-club organization. A study in mice found that a reduced diet with fewer proteins and calories led to a lower rich-club index, indicating that nutrition during development can influence structural connectivity (Barbeito-Andres et al., 2018). In humans, offspring of schizophrenia patients, but not of bipolar disorder patients, showed reduced rich-club organization for structural connectivity (Collin et al., 2017). Furthermore, structural covariance networks of patients showed disrupted rich-club organization, along with increased path length, for focal cortical dysplasia (type II), subcortical nodular heterotopia, and polymicrogyria (Hong et al., 2017).

An analytical expression, including a null model that allows for a quantitative discussion of the rich-club phenomenon, was described in Colizza et al. (2006). Within this framework, the rich-club coefficient is defined as

$$\phi(k) = \frac{2E_{>k}}{N_{>k}(N_{>k}-1)},$$

where $E_{>k}$ is the number of edges among the $N_{>k}$ nodes that have a higher degree than k and $N_{>k}(N_{>k}-1)$ is the maximum number of possible edges among the $N_{>k}$ nodes. Essentially, this coefficient measures the ratio of existing connections to the potential

Formation of Hubs

number of connections among a given set of nodes. Hence, the rich-club coefficient is a vector of values (different coefficients for different values of k).

In order to simplify the comparison between different complex networks, we can summarize the rich-club organization as the mean of the rich-club coefficients for all degrees k that qualify a node as being a hub (Bauer and Kaiser, 2017). We can therefore define a so-called hub-rich-club coefficient (HRCC) as:

$$HRCC = \sum_{k \in \Gamma} \phi(k) / |\Gamma|,$$

where Γ is the set of the degrees of all hub nodes in a given network.

As for the definition of hubs, there is no clear threshold for when to classify a network as "rich-club"; instead, containing rich-club features is a graded property.

8.3 Mechanisms for Generating Hubs

Preferential Attachment
In this approach for network generation, nodes that initially have a higher degree have a higher probability to receive links from nodes that are added to the network. This preferential attachment to already highly connected nodes is also called "rich-gets-richer" or "the Matthew effect." Starting with a small initial randomly connected network, further nodes are added one by one to the graph by preferential attachment. At the beginning of this process, the probability that a new node is connected to an existing node i is as follows:

$$P(i) = \frac{k_i}{\sum_j k_j},$$

where k_j is the number of connections of the node j (Barabasi and Albert, 1999). After establishing a connection to node i, the probabilities are recalculated to reflect the nature of scale-free networks: if i is connected to j, then it is more likely that i is connected to nodes that are already connected to j and it is less likely that i is connected to nodes that are not connected to j.

This approach results in networks with hubs and a power-law ("scale-free") degree distribution. While sparse networks can be generated straightaway, for a dense network where 10% or more of all potential connections exist and where the clustering coefficient is high, a modified approach is needed. Instead of starting with a random network, we need to start with a fully connected network and use rescaling for the probability $P(i)$ of a new node to connect to an existing node i (Kaiser, Martin, et al., 2007):

$$P^*(i) = \begin{cases} k_i\, P(i), \text{if } i \text{ and } i^* \text{ are connected} \\ P(i), \text{if } i \text{ and } i^* \text{ are not connected} \end{cases}$$

where

$$P(i) = \frac{P^*(i)}{\sum_j P^*(j)}.$$

Preferential attachment can generate a hub organization and robustness toward removing nodes or edges, as found in cat and macaque structural connectivity (Kaiser, Martin, et al., 2007). However, it assumes that connections of a new node form depending on how well-connected a potential target node is compared to all other existing nodes. The knowledge of such global information is often difficult to justify in a biological context; for example, axonal growth cones in neuronal networks can sense only their local environment. It is unclear whether and how the preferential attachment model can be applied to such network growth dynamics, as it often cannot provide a mechanistic explanation (Bauer and Kaiser, 2017).

Old-Gets-Richer
An alternative mechanism for the formation of hubs is the old-gets-richer model, based on the consideration that older nodes that are established early on during development have more time to receive connections from other nodes in the network. Young nodes, established at the end of development, on the other hand, arise at a time when most other nodes have already matured and are not establishing connections with the more recently generated nodes. Therefore, nodes that arise early have the potential to receive many connections and become network hubs (Varier and Kaiser, 2011).

Indeed, for *C. elegans*, neurons that are generated early during development tend to accumulate more connections and tend to be hubs of the adult network (Varier and Kaiser, 2011): all neurons that become hubs are born within 500 minutes, before the hatching event around 800 minutes (see figure 8.3A). In particular, the 11 hub nodes that form the *C. elegans* rich-club network are born early before the developmental elongation of its body (Towlson et al., 2013). For the macaque brain, regions of the archi- and paleocortex that mature earlier during brain development receive more incoming fiber tracts from younger brain regions of the neocortex (Kaiser, 2007). This effect, even though to a lower extent, is still visible between the early developing parietal and occipital lobes compared to the later developing regions of the neocortex (see figure 8.3B). It means that old nodes can receive incoming connections from all later nodes, leading to a higher overall degree. Younger nodes, on the other hand, receive

Formation of Hubs

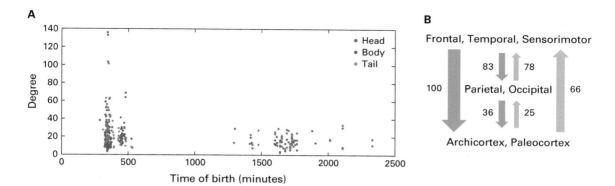

Figure 8.3
Old-gets-richer: older nodes, generated early during development, have a higher number of connections. (A) Node degree, the number of connections of a neuron, in relation to its time of origin. Note that all nodes with more than 33 edges originate early during development (after Varier and Kaiser, 2011). (B) For the macaque, fiber tracts follow the old-gets-richer model where regions that are ontogenetically (and phylogenetically) older (archi- and paleocortex) receive more incoming connections from younger nodes (neocortex) whereas younger nodes receive fewer incoming connections from older nodes (based on Kaiser, 2007).

fewer incoming connections from existing nodes as these finish connection establishment earlier, leading to a lower overall degree and a lower proportion of incoming connections.

Nonlinear (Accelerated) Growth

So far, we have looked at approaches where one node is added at each step. However, this is not realistic concerning the growth of biological systems. If we assume that neurons arise due to cell division during development, one node will divide into two nodes, these two nodes will divide into four nodes, these four nodes divide into eight nodes, and so on (see figure 8.4 for different growth models). That means that the number of new nodes that are generated during each step will increase over time. As a result, the total number of nodes in a network will not increase at a constant rate, one node at a time, leading to linear growth of network size. Instead, the total number of nodes will follow a nonlinear growth with more nodes being added at each step. As the number of new nodes increases, we will also call this accelerated growth (Bauer and Kaiser, 2017).

As an example of growth in biological systems, the number of cells per unit volume c in the growth phase of bacterial cultures can be described by an exponential function:

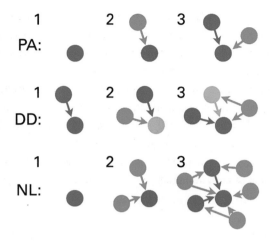

Figure 8.4
Growth models leading to highly connected nodes. PA, preferential attachment: New nodes (red) preferentially connect to nodes with higher degrees. DD, duplication–divergence model: At each step, a random node (light red) is duplicated (red) together with its links. NL, nonlinear growth: The number of new nodes that are added at each step increases nonlinearly over time. New nodes project to already present nodes (blue) establishing on average a connections (NL$_A$) or link to each existing node with a probability p (NL$_P$) (after Bauer and Kaiser, 2017).

$$c(n) = c_1 2^n$$

where c_1 is the initial number of cells of the culture, and n is the number of divisions a cell has undergone (Monod, 1949). For brain evolution, it was proposed that new neural structures form by separation of already existing areas (Ebbesson, 1980, 1984), with the number of brain areas then increasing exponentially. Work in brain evolution suggests that when new functional structures are formed by specialization of phylogenetically older parts, the new structures largely inherit the connectivity pattern of the parent structure (Ebbesson, 1980). This means that the patterns are repeated and small modifications are added during the evolutionary steps that can arise by duplication of existing areas (Krubitzer and Kahn, 2003). Such inheritance of connectivity by copying modules can lead to scale-free systems (Ravasz et al., 2002).

Duplication Divergence
For the duplication-divergence model, each time step consists of a duplication and a divergence step. In the duplication step, a random node is duplicated, that is, it projects to the same connection partners as the original node. The connection between the two

duplicated nodes prevails with probability p. In the subsequent divergence step, one of the two connections of the partners with the duplicated nodes is pruned with probability q. Maximally one of the two connections can be pruned, and so this model is based upon nontrivial coordination between the two connections.

The duplication-divergence model (Vázquez et al., 2003), which is inspired by the evolution of protein-protein interaction (PPI) networks, needs a duplication of previously existing connection patterns. Moreover, the divergence step entails a coordination between duplicated nodes (i.e., a new connection is pruned with a prespecified probability, but only if its counterpart is not). Such communication is difficult to justify in many biological networks, where the information exchange between maturing network components is limited.

Preferential Detachment

For the sake of completeness, we also mention the possibility that hubs arise at a late stage of development. While experimental evidence and previous models strongly suggest that hub nodes are formed early during development, we cannot exclude the option that *some* hubs form later on. If such a phenomenon occurs in neural systems, it would most likely happen during phases of network reorganization where connections between nodes can be strengthened or pruned away. Hubs would be unlikely to arise from random changes in connections so there would need to be a mechanism for preferential strengthening or detachment of connections. Indeed, we observed preferential detachment of structural connectivity in humans between the ages of 4 and 40 years (Lim et al., 2015). However, we did not find that this mechanism leads to a change in hub architecture of structural connectivity. Still, preferential mechanisms for hub formation might occur at different scales, for example, concerning the microconnectome, or for different types of connectivity, concerning hubs in functional or effective connectivity networks.

Networks resulting from different growth models can be compared with actual brain networks: axonal connections between brain regions in the rhesus monkey (Kaiser and Hilgetag, 2006) and the network between neurons in *C. elegans* (Varier and Kaiser, 2011; Varshney et al., 2011). The maturation time of a node (i.e., the time when the node is added to the growing network) is defined as the maturation time of a brain region during development for the macaque or the birth time of a neuron for *C. elegans*, respectively.

The nonlinear growth model (NL) assumes that the network size increases nonlinearly/exponentially with time (for code, see https://mitpress.mit.edu/changing-connectomes). At each developmental stage t, the network expands by d^t nodes (d is a parameter of the model) until a given network size is reached. The (rounded) number of newly formed nodes project to the already present nodes. Importantly, the formation of connections does not rely on any properties of the nodes. Hence, a newly developed node does not need to check whether its target node fulfills specific conditions, but forms connections dependent upon a model parameter that is initially specified. NL substantially differs from the well-known Erdös-Rényi model (ER; Erdös and Rényi, 1959), where the number of connections increases linearly, but the nodes are initially present already. We have compared the NL model with the extended preferential attachment (PA) and the duplication-divergence (DD) models (see figure 8.5).

Nonlinear growth can indeed produce networks with CV and HRCC values that are much larger than those of regular, random networks across different parameter values. As expected, these hub-related network measures are very dependent on the two model parameters: the nonlinear growth exponent d strongly affects CV values, by increasing the spread of the degree distribution. The NL_P model yields, because of equal probability of connections among all nodes, no rich-club organization. However, for $d > 1$ it produces hubs across a wide spectrum of parameters. For the NL_A model, the CV values and the rich-club organization express themselves much beyond what is expected based on comparable, regular networks.

The analyzed models reproduce many properties of the neural networks. However, only the NL_A model can account for all the data sets' CV and HRCC values. It provides an intuitive explanation for the origins of rich-club connectivity; during early network development, there are not many nodes to project to, and therefore most early born nodes project to each other. Since early developing nodes are likely to become highly connected, hubs are predominantly connected among each other. Furthermore, several models can reproduce the relative frequency of hub nodes that arise at different stages of development (see figure 8.6).

Overall, nonlinear growth relies solely on locally available information (e.g., nodes do not require knowledge about the degree of connection partners) and so can provide a baseline benchmark for modeling network evolution that contains phases of exponential growth.

In summary, we have looked at the formation of hubs including factors such as the birth time of nodes (old-gets-richer), the network growth (accelerated growth), and the degree of existing nodes (preferential attachment). Furthermore, hubs may also arise through modifications at a later stage such as the preferential removal of connections (preferential detachment). While the time for connection formation and the spatial embedding are important factors, they may not always be sufficient to explain the formation of hubs. Recent studies indicate that there are also genetic factors: using

Formation of Hubs

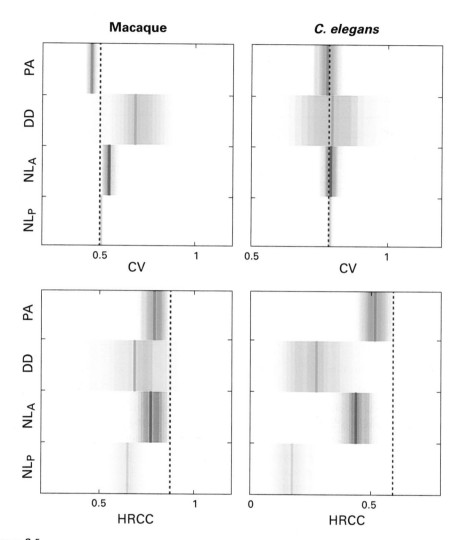

Figure 8.5
Generation of network properties. Degree variation (CV) and mean hub-rich-club coefficient (HRCC) of the models in relation to values of real-world networks (black dashed lines). Red, green, blue and cyan shaded regions indicate the distribution of values that preferential attachment (PA), duplication-divergence (DD), nonlinear growth with new nodes establishing a fixed absolute number of connections with existing nodes (NL_A), and nonlinear growth with new nodes establishing connections with a certain probability so that more connections are established when there are a larger number of existing nodes (NL_P) could yield, respectively. Thick colored lines indicate the mean values of the distributions (after Bauer and Kaiser, 2017).

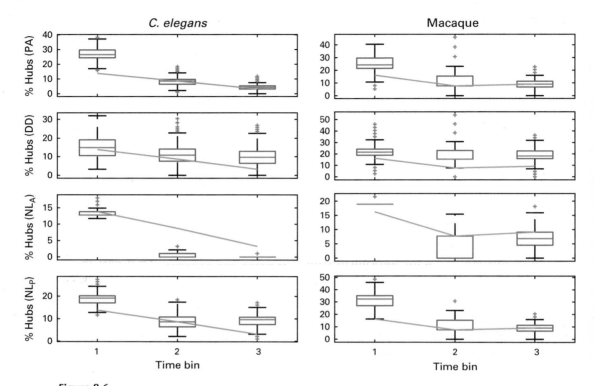

Figure 8.6
Trajectories of hub occurrences for *C. elegans* and macaque. Green lines indicate the experimental data, while the box-plots show the results of the (optimized) model-generated samples. The green lines are identical for the same data set but are displayed within different axes. PA, preferential attachment; DD, duplication-divergence; NL_A: nonlinear growth with new nodes establishing a fixed absolute number of connections with existing nodes; NL_P: nonlinear growth with new nodes establishing connections with a certain probability so that more connections are established when there are a larger number of existing nodes (after Bauer and Kaiser, 2017).

brain-wide gene expression atlases, characteristic gene expression patterns are associated with network hubs and are conserved across species and scales (Arnatkeviciute et al., 2019). A detailed analysis of the involved genes suggests that the genes involved in the development of hubs, such as the ones for myelination and synaptic transmission, are different from the genes with a metabolic role which signify hubs at the mature stage. Future studies will need to systematically assess the role of different factors through simulations of brain network growth and through comparisons with experimental data of network organizations at different developmental stages.

9 Module Formation

9.1 Overview

Early brain connectivity development consists of sequential stages: birth of neurons, their migration, and the subsequent growth of axons and dendrites. Each stage occurs within a certain period of time depending on types of neurons and cortical layers. However, also within each stage there is a sequential development with some neurons establishing connections earlier than others. When a group of neurons connects with other neurons only within a certain time during development, we can call this a *time window* for forming connections.

Neurons with the same time window are often, but not always, spatially nearby one another. In addition, neurons within a region tend to be born around the same time and often have the same cell lineage, inheriting internal genetic factors from their ancestors. Therefore, being born during the same time window influences both genetic predisposition and environment—that means the "desire" to connect, guided by genetic factors, and the "ability" to connect, based on the available neurons and regions at that point in time. For example, pyramidal cells that are part of the same cell lineage are more often connected than cells with a different developmental history (Perin et al., 2011).

Neurons within the same layer, the same column, or the same region tend to establish connections more often between themselves than outside their communities. For example, in the visual cortex in primates, 60% to 70% of all excitatory synapses arise from neurons within the same brain region (Young, 2000). This directly results in a modular architecture with many links within communities or modules and fewer to the rest of the network. Such modules tend to share a similar spatial location, a similar origin (cell lineage), and a similar time for their establishment. In this chapter, we will more closely look at the link between the timing of connection formation and the establishment of topological network modules.

9.2 Modules and Time Windows for Brain Networks

Time Windows during Brain Maturation

Brain development occurs at different time periods depending on regions, cell types, and types of development (Andersen, 2003; Rakic, 2002; Shaw et al., 2008). Cells are born, differentiate, migrate to certain regions of the brain, and form synaptic connections influenced by the aforementioned factors. The change of brain organization over time, with distinct trajectories for different parts of the brain, has been termed *chronoarchitecture* (Bayer and Altman, 1991). Initial overproductions of neurons and synapses are reduced after one year from birth, suggesting particular time periods for neurogenesis, programmed apoptosis, early synaptic pruning, and synaptogenesis (Purves and Lichtman, 1980; Huttenlocher, 1984; Rakic et al., 1986; Kelsch et al., 2010).

A detailed analysis of ten cortical regions found that while neurogenesis in primates starts almost at the same time, 38 to 40 days after conception (E38–40), time windows close between E70 for the limbic cortex and E102 for the visual cortex (Rakic, 2002). Therefore, as already mentioned in the previous chapter when discussing the formation of network hubs, neurogenesis and the formation of layers II through VI in some regions finishes several weeks earlier than in others. Note that layer I is an exception as, for all regions, neurons are added throughout the entire duration of corticogenesis.

The relative difference in birth time between a pair of neurons, independent of whether both neurons are born early or late, can influence connectivity. Two neurons that arise at the same time, through cell division of a common progenitor cell, have several shared features: their genetic mark-up, a nearby spatial location, and a relatively nearby starting time for axo- and synaptogenesis. Note that the starting time of axogenesis, relative to the birth time of neurons that form these axons, can vary for different cortical layers (Barone et al., 1996). Pyramidal cells that are part of the same cell lineage are more often connected than cells with a different developmental history (Perin et al., 2011). Even if two cells have different direct ancestors, being born at the same time still means that the formation of connections occurs concurrently unless one of the cells is involved in a cell migration process. For pairs of connected neurons in *C. elegans*, most are born within 50 minutes of each other while hatching takes place around 840 minutes (Varier and Kaiser, 2011).

Functional Effects of Time Windows: The Rise of Critical Periods

Critical periods are times during development when a skill or characteristic can easily be acquired which, outside those periods, becomes difficult or impossible to acquire (Hensch, 2004). Such periods can occur after birth, for example, as identified by Konrad Lorenz (Lorenz, 1958), the imprinting process by which birds that leave their nest early bond instinctively with the first moving object that they see within the first hours of hatching.

Focusing on brain development before birth, ocular dominance columns have been an example of the formation of modular circuitry within regions. Ocular dominance in primates and carnivores results from the segregation of LGN axons into eye-specific columns in layer 4 of the primary visual cortex (Hensch, 2004). These columns or modules arise during development following an initial stage where afferents from the LGN overlap extensively, a process that takes several weeks. At the same time, the refinement of these circuits critically depends on visual input: preventing normal retinal activity results in disrupted segregation in that ocular dominance columns do not form. Note, however, that the formation of ocular dominance columns starts much earlier than during the critical period. Findings in cats, monkeys, and ferrets indicate that columns develop far earlier, more rapidly, and with considerably greater precision than was previously suspected.

These observations indicate that the initial establishment of cortical functional architecture, and its subsequent plasticity during the critical period, are distinct developmental phases that might reflect distinct mechanisms (Katz and Crowley, 2002). Altogether, this highlights that distinct time windows are related to the formation of functional circuits, opening the possibility that changes in event timings can be linked to neurodevelopmental disorders as we will see in part III.

Links between Time Windows and Module Formation

The role of time windows for module formation has been experimentally observed in the fruit fly *Drosophila* (Chiang et al., 2011): neurons linking to other neurons in the same module are born around the same time with an early time window corresponding to the locomotor, an intermediate time window corresponding to the visual and olfactory, and a late time window corresponding to the auditory module.

Some studies have also reported preferential electrical coupling between neurons sharing genetic lineage that are likely to have similar developmental time windows in mice neocortex (Yu et al., 2009; Yu et al., 2012). Furthermore, nonoverlapping time windows among neurons in mouse CA3 resulted in selective synaptic connectivity forming submodules in the hippocampus (Deguchi et al., 2011; Druckmann et al., 2014).

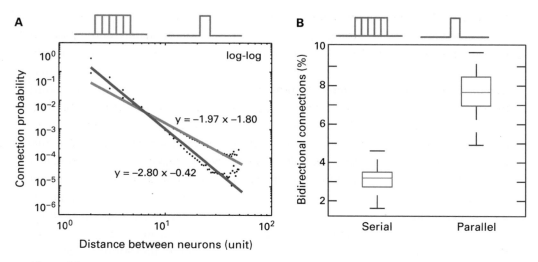

Figure 9.1
Formation of long-distance and imbalanced connections. (A) Connection length distribution for simulated sequential (red, no overlap between time windows) and parallel (blue, identical time window for all neurons) network growth of 1,400 neurons distributed in three-dimensional space. (B) Boxplot of the percentages of reciprocal connections for serial and parallel growth across 50 simulation runs (based on Lim and Kaiser, 2015).

9.3 Mechanisms for Generating Modules

Sequential versus Parallel Connection Formation

Before we look at separate time windows, we can have a first look at timing in general. For this, we observe the extreme cases of network development to investigate the effects of birth times, developmental time windows, and competition between neurons (Kaiser et al., 2009), where all nodes are either starting to form connections at the same time (see figure 9.1, parallel case: overlapping time windows, blue) or where they form connections sequentially, one node after another (see figure 9.1, serial case: nonoverlapping time windows, red) (Lim and Kaiser, 2015). Sequential growth leads to a higher proportion of long-distance connections (see figure 9.1A): As all previously established neurons have finished the connection formation, fewer "docking spaces" (Kaiser et al., 2009) at target neurons are available so that axonal growth cones need to travel further to find a target neuron with free space on the soma or dendritic tree. On the other hand, for parallel growth where connection formation is ongoing for most neurons, many free places are still available, leading to shorter axon lengths for established connections. This easier connection formation during parallel growth results in more bidirectional connections between pairs of neurons due to overlapping developmental time windows (see figure 9.1B).

Comparison with *C. elegans* Network Formation

How do those predictions compare with experimental data? We tested our model with the only connectome data set where both neuronal connectivity and birth time, as an estimate for the lower bound of starting to form connections, exists, that is, the connectome of *C. elegans*, focusing on connections formed by chemical synapses (Choe et al., 2004; Chen et al., 2006; Hall and Altun, 2008; Varshney et al., 2011). We tested our three major predictions from our model concerning degree, connection lengths, and bidirectional connectivity. Groups 1, 2, and 3 represent three groups of neurons clustered based on their birth times: neurons in group 1 and group 2 have similar birth times and were born early, whereas neurons in group 3 were born much later compared to those in groups 1 and 2 (see figure 9.2A). While birth times in groups 1 and 2 were, on average, 114.78 minutes apart, the time difference between groups 2 and 3 was more than 1,255.70 minutes. As birth times of group 1 and group 2 do not differ much, we can assume that group 1 and group 2 represent the case of large overlapping time windows (or parallel growth), while group 1 and group 3 or group 2 and group 3 indicate the case of small overlapping time windows (or serial growth). Here birth time is used as an equivalent of the starting point of the time window for axon (and dendrite) outgrowth. Normally, time windows for neurogenesis and synaptogenesis should be treated differently; however, here we assumed time windows for synaptogenesis in *C. elegans* started after equivalent time passes for all neurons for simplicity.

Earlier-born neurons in *C. elegans* acquired higher degree (see figure 9.2B), longer axon lengths (see figure 9.2C), and higher reciprocal connectivity (see figure 9.2D), which were consistent with the model predictions (Lim and Kaiser, 2015).

Note that the size of the embedding space of neurons for axon growth and synaptogenesis of our model was fixed during development, while internal volume changes through neurite growth and external mechanical factors could change the location of neurons and influence their synapse formation probabilities. For uniform expansion along all directions, this would increase connection lengths but differences between serial and parallel growth would remain.

Multiple Time Windows

Similar time windows are due to a comparable cell lineage from common progenitor cells or common formation times of brain regions. As neurons or brain regions arise

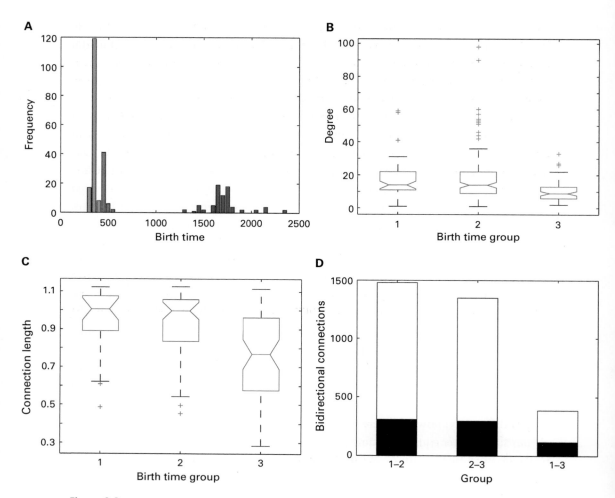

Figure 9.2
Model predictions versus neuronal connectivity in *C. elegans*. (A) Group 1 (green), 2 (red) and 3 (blue) using *k*-means clustering based on the birth times of neurons in *C. elegans*. x-axis: birth time (minutes); y-axis: number of neurons born within each birth time window. (B) Birth time group and degree of neurons. (C) Birth time group versus connection lengths (mm), estimated by Euclidean distances at the adult stage, of long-range connections. (D) Birth time group and the number of bidirectional connections. x-axis: birth time groups; y-axis: the number of connections (black: bidirectional connections; white: the total number of connections between relevant groups) (after Lim and Kaiser, 2015).

during division of earlier neurons or regions, they also tend to be spatially nearby one another. We can therefore think of similar time windows (see figure 9.3A) for nodes within the same spatial cluster, following a similar cell lineage (see figure 9.3B). If two nodes with the same time window for sending and receiving connections are more likely to connect to each other, a network where each node can have one out of three time windows results in a network with three network modules (see figure 9.3C, D).

For this approach (for code, see https://mitpress.mit.edu/changing-connectomes), the establishment of an edge depends on the distance between the nodes and the current likelihood of establishing a connection given by the time windows of both nodes. The distance-dependent probability is

$$P_{\text{dist}}(d) = \beta e^{-\gamma d}$$

where d is the spatial Euclidean distance between two nodes, $\gamma = 6$, and $\beta = 6$.

The time-dependent probability P_{time} of a node is influenced by its distance to pioneer nodes, one for each time window. These pioneer nodes are the basis for later network clusters. Each node, belonging to time window (i), has a preferred time for connection establishment, and the probability decays with the temporal distance to that time:

$$P_{\text{time}}^{(i)}(t) = P(t, \mu^{(i)}, \Sigma_{\mu^{(i)}}(\alpha)) = \frac{1}{16}(t^{2\lambda}(t^\lambda - 1)^2)^{\frac{1}{\Sigma_{\mu^{(i)}}(\alpha)}}$$

with

$$\mu^{(i)} = \frac{i}{k+1},$$

with k being the total number of time windows and

$$\lambda = -\frac{\log(2)}{\log(\mu)}$$

and α as the desired value of the integral so that $\int_0^1 P_{\text{time}}^{(i)}(t)dt = \alpha$ and $\Sigma_{\mu^{(i)}}(\alpha)$ is a numerically determined scaling factor to get the desired integral value α.

The probability P that two nodes U and V will connect then depends on both the distance between them as evaluated by P_{dist} and the time windows of connection formation of nodes U and V:

$$P = P_{\text{dist}}(d(U,V)) \cdot P_{\text{time}}^{(w(V))}(t) \cdot P_{\text{time}}^{(w(U))}(t)$$

The size of a module is determined by the width of the time window while the amount of connections between modules is determined by the overlap of the time

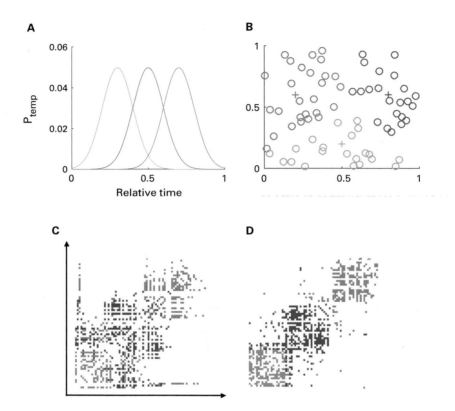

Figure 9.3
Network growth with three time windows. (A) Temporal dependence P_{temp} of projection establishment depending on node domain. Relative time was normalized such that zero stands for the beginning of development and one for the end of network growth. The three seed nodes had different time windows which were partially overlapping. (B) Two-dimensional projection of the 73 three-dimensional node positions. The color coding represents the time window corresponding to one of the three seed nodes (+). (C) Timed adjacency matrix (the first nodes are in the left lower corner). (D) Clustered adjacency matrix. The matrix is the same as in (C), but nodes with similar connections are arranged more adjacent in the node ordering (after Kaiser and Hilgetag, 2007).

windows (Kaiser and Hilgetag, 2007; Nisbach and Kaiser, 2007). We can calculate the overlap Ov between two time windows (*i*) and (*j*) as the overlap between their time window functions $P^{(i)}$ and $P^{(j)}$ as

$$\mathrm{Ov}(P^{(i)}, P^{(j)}) = \frac{\int_0^1 P^{(i)}(t) P^{(j)}(t) dt}{\sqrt{\int_0^1 (P^{(i)}(t))^2 dt \cdot \int_0^1 (P^{(j)}(t))^2 dt}} = \frac{\langle P^{(i)}, P^{(j)} \rangle}{\|P^{(i)}\|_2 \cdot \|P^{(j)}\|_2}$$

where $\langle \cdot, \cdot \rangle$ is the standard scalar product on the Euclidean vector space of integrable real valued functions on the interval [0, 1] and $\|\cdot\|_2$ is the associated norm.

While the number of modules in general corresponds to the number of unique time windows, a large number of time windows leads to a larger overlap between time windows. For this case, connections between modules become so frequent that modules start to merge, and the number of detected modules becomes lower than the number of time windows (Nisbach and Kaiser, 2007). Note that, in addition to the overlap between the time windows, the number of connections between modules also depends on how spatially close the initial seed nodes of both modules are. As connections preferentially form to nearby nodes, spatially adjacent clusters of nodes will end up with more connections between them.

In addition to a modular architecture, the generated networks also exhibit features of small-world networks (Nisbach and Kaiser, 2007). The degree distribution, however, follows an exponential distribution indicating that separate mechanisms, discussed in the earlier chapter, are needed to generate network hubs.

Reproducing connectivity across species

An alternative model for the formation of modules uses a neural activity model where the system is optimized to reduce wiring length and increase performance, resulting in a number of modules that corresponds to the number of optimized tasks (Jacobs and Jordan, 1992). However, such a mechanism would operate on an already formed network and might not be needed if modules have already been established through the time window mechanism discussed above.

Indeed, a related temporal approach, without the use of neuronal activity, could already lead to similar structural connectomes as observed in *Drosophila*, mouse, macaque, and human (Goulas et al., 2019). The characteristic that different parts of the network develop at different times is here termed *heterochronicity* in contrast to *tautochronicity*, where all network components arise at the same time. The model consists of spatially embedded and heterochronous neurogenetic gradients, without axonal guidance molecules or activity-dependent plasticity (see figure 9.4).

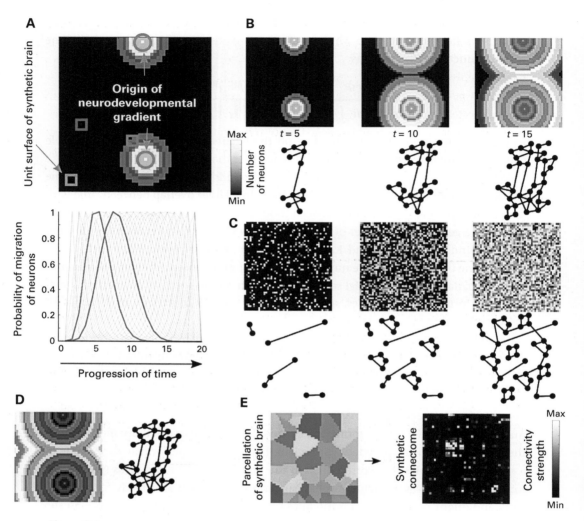

Figure 9.4

Developmental modeling approach. (A) Neurogenetic gradients were simulated in a synthetic 2-D brain. Each surface unit was characterized by a time window that indicates at a given time point the probability of a neuronal population migrating to each surface unit. Each time window is shaped by the distance of each surface unit from the root(s), that is, origin(s), of the neurogenetic gradients. For instance, the surface unit close to the neurogenetic origin (petrol green) is more probable to be populated earlier than the surface unit further from the neurogenetic origin (magenta). (B) Heterochronous and spatially ordered ontogeny of synthetic connectomes. (C) Heterochronous, spatially random and (D) tautochronous ontogeny of synthetic connectomes. (E) Creation of the synthetic connectome (after Goulas et al., 2019).

Only heterochronous development was able to reproduce the topological features, such as modularity and short path lengths, that were observed in the diverse connectomes, from *Drosophila* to humans. In order to do so, gradients of development need to be heterochronous and spatially ordered, rather than random.

Note that several other models have been proposed for module formation: generative models to observe changes across the life span (Betzel et al., 2016), topological reinforcement where Hebbian learning strengthens connections between neurons with high neighborhood similarity (Damicelli et al., 2019), a neural-mass model with strengthening of connections due to synchronous activity and homeostatic distance-dependent outgrowth of new connections between regions (Stam et al., 2010), and an interaction of neurite outgrowth, cell migration, and activity in modeled cell cultures (Okujeni and Egert, 2019).

Module refinement

The above mechanisms show that time windows, or heterochronous development, can generate multiple network modules. However, the network observed at the adult stage might differ from these generated networks because of processes that occur during or after initial module formation. Imaging studies have reported systematic macroscopic pruning of connections in the brain during development, which was not random but rather selective, similar to synaptic elimination (Supekar et al., 2009; Homae et al., 2010; Gao et al., 2011; Lim et al., 2015). Baby connectome studies showed that modularity and clustering coefficient increased from preterm to term babies (Tymofiyeva et al., 2013; van den Heuvel et al., 2015), whereas modularity and clustering coefficient decrease during adolescence and adulthood, leading to a more distributed and integrated network (Fair et al., 2009; Hagmann et al., 2010), which may result from eliminating diffusive, less accurate, and more redundant connections (see chapter 11 for more details).

In summary, several mechanisms relating to spatial and temporal vicinity, often related to similar cell lineage, can result in an early modular network organization. Further refinement through activity-based plasticity and removal of cells (apoptosis) or connections can lead to the modular organization that is observed after connectome maturation.

10 Cortical Folding

10.1 Overview

Increasing brain size and number of neurons at a higher rate than body size increases cognitive abilities (Herculano-Houzel, 2017). However, there are limits to brain expansion. One could think of metabolic limits in terms of powering brains: for humans, the brain weighs around 2% of the body mass but consumes 20% of the energy. A more severe limit for brain size is given by constraints on the skull and head size. During birth, the head has to pass between the pubic bones through the birth channel. One option, as used for humans and other primates, to still enlarge brain size is to move large sections of brain growth and maturation to the time after birth. Another option, used for humans and many other species, is to increase the number of neurons and gray matter surface area through cortical folding. By introducing "wrinkles" into the surface, with gyri (singular: gyrus) on the outside and sulci (singular: sulcus) inside these folds, the total size of the surface area can be increased. For humans, around 60% to 66% of the cortical surface is "hidden" inside sulci (Wang et al., 2016) while this ratio is 59% for the macaque (Van Essen and Drury, 1997). Folding can vary between different primates: whereas rhesus monkeys show folding (gyrencephalic brains), marmoset monkeys do not (lissencephalic brains). However, for 34 primate species, gyrencephalic primate neocortices exhibited a stable fold wavelength of about 12 mm despite a 20-fold variation in cerebral volume (Heuer et al., 2019). Also within humans, there is a wide variation in that the largest brains can have up to 20% more surface than a scaled-up small brain because of increased gyrification (Toro et al., 2008).

What does folding have to do with connectome development? Briefly, the organization of fiber tracts is linked to the folding pattern of the brain, and, according to some theories, there are mechanistic relations between fibers and the formation of wrinkles. In addition, changes in brain development, as observed in neurodevelopmental diseases, often coincide with changes in cortical morphology, including changes in folding.

10.2 Cortical Folding

The Timing of Folding

During early development, neural progenitor cells undergo a symmetrical cell division, leading to exponential growth in the number of cells. In humans, after 6 weeks of fetal life, neurons start to undergo asymmetrical divisions with one cell staying in the ventricular zone and continuing cell division and the other cell maturing into a neuron and starting migration. At this stage, neurons of the ventricular zone migrate along radial glia cells to the outer layers of the brain, forming the emerging cortical gray matter.

The radial-unit hypothesis by Pasko Rakic (Rakic, 1995) proposes that the surface area of the brain is linked to the number of radial units formed by symmetrical division within the ventricular zone within the first fetal weeks. Small changes at this stage can have dramatic effects on the later layer thickness or cortical gyrification: for example, one additional symmetrical cell cycle could double the amount of progenitor cells, leading to more migrating cells (White et al., 2010). Due to exponential growth, structures that develop for a longer time will develop a larger surface area, leading to the phrase "late equals large." Indeed, such a pattern has been observed when timing and region size are looked at across many species (Striedter, 2005).

At later stages apoptosis, resulting in the elimination of up to 50% of initially generated neurons (Cowan et al., 1984), can influence cortical layer formation, as we saw in chapter 6, and thus cortical thickness. In turn, it might influence the folding pattern during brain development.

Theories

Given the growth of gray matter volume and surface area, why is the cortex folded? Why are brains in some species folded but not in others? And, finally, why does the cortical folding pattern differ between individuals, making it as unique as a fingerprint?

Models for cortical folding can be divided into studying the influence of gray matter and white matter forces (see figure 10.1). White matter theories claim that the cerebral cortex is folded because of mechanical forces of *axonal tension* exerted by the underlying white matter fibers (Van Essen, 1997; Seldon, 2005; Nie et al., 2012; Mota and Herculano-Houzel, 2012, 2015). The "pulling theory" suggests that white matter axons connecting two brain regions exert a pulling force that brings these two regions closer

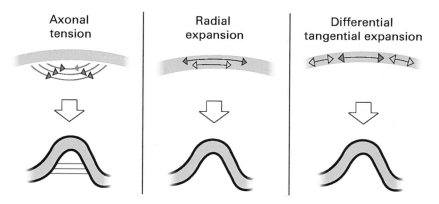

Figure 10.1
Schematic illustration of different folding theories. Axonal tension: Fiber tracts between regions cause a tension leading to folding when the surface between regions expands. Radial expansion: Upper and lower cortical layers expand at different rates (in this case, the upper layers expand faster) leading to folding. Differential tangential expansion: Different regions show a different expansion rate (adapted from an image provided by Yujiang Wang).

together, resulting in cortical folds (Van Essen, 1997). This is supported by tract-tracing experiments (Hilgetag and Barbas, 2006), which demonstrate that dense white matter fiber bundles are mostly straight. However, the empirically observed direction of fiber tensions appears inconsistent with that predicted by the pulling theory (Xu et al., 2010). In contrast, the "pushing theory" suggests that white matter axon bundles push the gray matter from below, leading to the formation of gyri (Nie et al., 2012; Chen et al., 2013). This is supported in that white matter streamlines inferred from diffusion imaging mostly terminate in gyri. However, this empirical observation may be driven by a technical limitation of diffusion imaging, leading to what is termed a gyral bias (Ronan and Fletcher, 2015). Moreover, there are several lines of evidence against white matter theories. First, folding occurs before axonal innervation. Second, reducing the amount of axonal connections to the cortex, for example, shown in enucleation experiments, leads to an increase in gyrification. Third, the pattern of tension caused by fiber tracts is inconsistent with the pattern of folding.

The rival gray matter theories propose that cortical gyrification arises because of expansion and deformation of the gray matter itself (Nie et al., 2010; Bayly et al., 2013; Tallinen et al., 2014; Ronan and Fletcher, 2015). *Radial expansion* theory proposes that upper and lower cortical layers, when expanding at different rates, can lead to a bulging of the surface. However, cells which contribute to supragranular (upper layer) expansion do not differ between lissencephalic and gyrencephalic species. Therefore, this points to a primary role of expansion without the need for a differential between upper and lower layer expansion. Instead, *differential tangential expansion* of the gray matter, where different surface regions expand at different rates, might underlie the pattern specificity of gyrification (Ronan and Fletcher, 2015). Comparisons between species has shown that prolonged neurogenesis, an increase in the number and types of progenitor cells, and conical migration trajectories increase gyrification even though none of those factors are unique for gyrencephalic species. Finally, genes that promote neuronal differentiation, apoptosis, or radial migration reduce gyrification in contrast to genes that increase the proliferation of progenitor cells (Ronan and Fletcher, 2015).

Note that gyrification might be caused by a combination of these different models. Moreover, the gyrification at the mature stage might result from different factors occurring at different times as well as secondary gyrification changes due to apoptosis.

Local Changes due to Folding

Forces generated by cortical folding influence laminar morphology and appear to have a previously unsuspected impact on cellular migration during cortical development (Garcia et al., 2018). In line with the geometrical predictions for an isometric (volume-preserving) folding of the cortical sheet, a correlational analysis showed that the relative thickness of the upper cortical layers was collectively reduced in gyri and increased in sulci in macaque monkeys. Reverse relations were obtained for the deep layers (Welker, 1990; Hilgetag and Barbas, 2006).

Changes with Age and Gender

Gyrification has been shown to decrease with aging. During adolescence, there are widespread reductions in gyrification in cortical regions including precentral, temporal, and frontal areas. These decreases in gyrification only partially overlap with changes in thickness, volume, and surface of gray matter (Klein et al., 2014).

In addition to age, there are differences between sexes. While these differences do not appear at the global level, observing gyrification of the whole brain, they were found when parcellating the cortex into several regions (Luders et al., 2004) or using a local measure of gyrification (Luders et al., 2006).

Changes with Disease

Aberrant gyrification was described in several pathological neurodevelopmental conditions (Hedderich et al., 2019). For example, a decrease in gyrification was found in attention deficit hyperactivity disorder (ADHD) and dyslexia, while increased gyrification was described in Williams syndrome, autism spectrum disorders (Hardan et al., 2004; Nordahl et al., 2007), and schizophrenia (Sallet et al., 2003; Harris et al., 2004; Csernansky et al., 2008). Abnormal folding patterns also occur for certain types of epilepsy (Voets et al., 2011) and other neurodevelopmental disorders (Pang et al., 2008). Note that changes in brain size will affect global changes while more specific factors would be linked to local changes.

Changes with Learning and Nutrition
Changes in behavior can also affect cortical folding. Meditation practitioners show increased cortical gyrification in several regions with increases for the right anterior dorsal insula being linked to the years of meditation practice (Luders et al., 2012). Concerning nutrition, a study on anorexia demonstrated that reduced body weight has an impact on cortex morphology: in anorexic patients, gyrification was significantly decreased but changed back to normal after body weight restoration (Bernardoni et al., 2018). This raises the important point that reduced gyrification as observed for psychiatric disorders, such as schizophrenia or depression, might be due to weight loss caused by antipsychotic medication or alterations in nutrition.

10.3 Mechanisms for Generating Folding

Measuring Folding
To measure folding, the cortical surface area and its shape have to be determined. Common measures of folding include the following:

Gyrification index (GI). The gyrification index is the ratio between the total surface area, including both gyri and sulci, and the exposed surface area (see figure 10.2). The exposed surface can be thought of as the surface area that one would obtain if the brain were wrapped in cling film. The total surface corresponds to the pial surface as reconstructed through neuroimaging tools in the case of MRI data.

Curvature. Following a triangulation of the cortical surface, the curvature for each vertex of a triangle can be calculated. In this way we can estimate the minimum and maximum curvature for each triangle. The average of these two measures gives the mean curvature (or extrinsic curvature). The multiplication of these two measures gives the Gaussian curvature (or intrinsic curvature). The curvedness is measured by the square root of the sum of the squares of the maximum and minimum curvature. The intrinsic curvature of the cortex can predict the degree of gyrification in different regions of the brain across healthy subjects (Ronan and Fletcher, 2015).

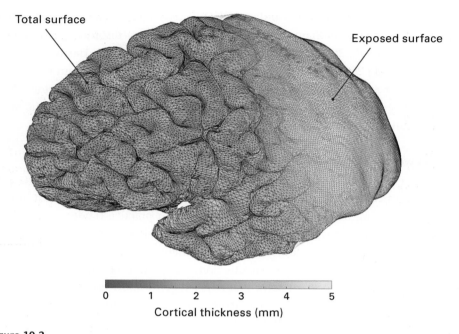

Figure 10.2
Assessing cortical folding (gyrification). The gyrification index is the ratio of the total gray matter surface area over the exposed gray matter surface area shown here for one human subject (the variation in cortical thickness is also shown). The gyrification index for subjects in the Human Connectome Project data set, ages 22 to 25 years, is between 2.5 and 3 (Wang et al., 2016). (Figure by Yujiang Wang, reproduced with permission from http://xaphire.de/page3/.)

Intrinsic curvature index (ICI). Integrates across all regions of positive intrinsic curvature and divides by 4π (the integrated intrinsic curvature for a perfect sphere of any size). ICI is calculated as follows:

$$ICI = \frac{1}{4\pi} \iint k' dA,$$

where $k' = | k_{max} k_{min} |$ if $k_{max} k_{min} > 0$, or else $k' = 0$. Excluding regions of negative intrinsic curvature ensures that the spherical component of each dimple or bulge is not canceled by the saddle-shaped zone around its perimeter. Any local dimple or bulge having the shape of a half-sphere increments the intrinsic curvature index by a value of 0.5, independent of its size (Van Essen and Drury, 1997). Note that the integration equation above assumes a smooth curvature; for a triangulated surface reconstruction, a discrete version of this approach would be used.

Folding index (FI). Integrating the product of the maximum principal curvature and the difference between maximum and minimum curvature, *FI* is defined as follows:

$$FI = \frac{1}{4\pi} \iint |k_{max}|(|k_{max}| - |k_{min}|)dA,$$

where A is the cortical surface area and k_{max} and k_{min} are maximum and minimum curvature, respectively (Van Essen and Drury, 1997). For a sphere, *FI* = 0 whereas *FI* > 0 for an ellipsoid.

Fractional dimensionality. An alternative measure to assess the shape of the cortex has been proposed under the name fractional dimensionality (for a recent application to calculate cortical fractional dimensionality, see Madan and Kensinger, 2016). Fractional dimensionality, calculated using a box-counting algorithm or a dilation algorithm, measures the complexity of the surface shape. For human data, fractal dimensionality is more sensitive to age-related differences than either cortical thickness or gyrification index (Madan and Kensinger, 2016).

Mechanical Models

To understand how changes in the growth or death of cells and connections can alter brain morphology, we need to understand how components of the brain are physically interacting. Both gray and white matter theories rely on physical forces that lead to the deformation of the cortical surface leading to gyrification. Some of these physical principles are (after Garcia et al., 2018) as follows:

Force equilibrium. The net force on the tissue is equal to the product of its mass and acceleration. As acceleration for the slow process of gyrification is almost zero, we can assume perfect force equilibrium (zero net force).

Stress. Stress is force normalized by the area over which the force is applied. Pressure is an example of stress where the direction of the force is normal to the surface area. Shear stresses, on the other hand, describe forces tangent to a surface. Tensile stresses elongate while compressive stresses shorten the tissue.

Strain. Strain describes the deformation of the tissue normalized by its size. Axial strain is the (positive) change in length divided by the original length.

Material properties. Material properties describe to what extent the applied forces can lead to deformation of the tissue. Brain tissue is widely understood to be viscoelastic, such that sustained tension can lead to stress relaxation as the tissue lengthens passively: elastic stretch (tension) leads to a lengthened configuration with no stress while compression leads to a shortened configuration with no stress.

Mechanical feedback. Tissue properties can influence and be influenced by environmental forces. For example, soma and axon shape are influenced by the cytoskeleton formed by microtubules and actin filaments. Furthermore, an increase in cell number and cell density (cells per mm^3) can influence the response to external forces. At the same time, applied forces can influence gene expression and actin polymerization.

Physical Models

While cortical folding is usually observed in computer models, it is also possible to look at tissue growth in a physical model. For this, 3-D printed layered gel can take on the same shape as a smooth fetal brain as measured through MRI (Tallinen et al., 2016). When immersed in a solvent, the outer layer swells relative to the core, mimicking cortical growth. The mechanical pressures between the expanding outer layer and the constrained core lead to the formation of sulci and gyri as found in later stages of the fetal brain. This behavior of the physical model also matches the characteristic patterns seen in a computer simulation of soft tissue with a growing cortex. In a sense, this approach is an interface model that combines gray and white matter theories as the difference in the tissue properties of both is driving gyrification. However, a closer comparison with experimental data, also including structural connectivity and folding at different developmental stages, will be needed to evaluate the contribution of this model to the observed gyrification.

Scaling Laws

As we saw above, an increased growth of outer layers (the gray matter) surface area can cause cortical folding. The relationship between surface area growth and the increased number of progenitor cells can be described through scaling laws. Given the total cortical surface area A_G, the exposed cortical surface area A_E, and the average cortical thickness T, there is a power law in that

$$T^{1/2} A_G = k A_E^{5/4}$$

(Mota and Herculano-Houzel, 2015). The only free parameter is k, or offset, a dimensionless coefficient that is presumed to be related to both the axonal tension and the pressure of cerebral spinal fluid.

It follows that for lissencephalic species, where A_G equals A_E, $T = k^2 A_G^{1/2}$. If T assumes lower values, the brain of species would be predicted to be gyrencephalic. Experimental data largely follows these predictions, indicating that cortical folding scales universally across clades and species, implying a single conserved mechanism throughout evolution (Mota and Herculano-Houzel, 2015). This single universal relationship for cortical

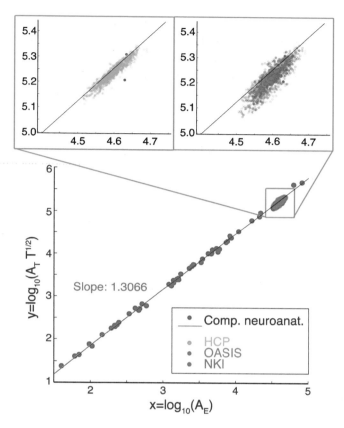

Figure 10.3
Comparing human data sets derived from magnetic resonance imaging (MRI) to *ex vivo* mammalian brain data. The scaling law for the comparative neuroanatomy data across different species is shown in gray. The gray regression line is also obtained for this data, with a slope of 1.307. Overlaid are the human MRI-derived data sets: green for Human Connectome Project (HCP), blue for Open Access Series of Imaging Studies (OASIS), and magenta for Nathan Kline Institute (NKI). The insets (red frame) show the human MRI-derived data in more detail relative to the regression line of the interspecies data. Note the two gray dots in the zoom-ins are from the previous comparative neuroanatomy data set for humans (after Wang et al., 2016).

expansion means that there is a transition point between smooth and folded cortices: Gyrencephaly ensues when A_G expands faster than T^2 (Mota and Herculano-Houzel, 2015). This typically happens when the cortical surface area A_G reaches 400 mm².

This relationship also holds within species and for different brain regions, as tested for human brains (Wang et al., 2016; Wang et al., 2019). Adult healthy human cortices, with gyrification and surface area estimated from MRI data, conform to the same scaling relationship (see figure 10.3). With age, the value of *k* decreases monotonically, and it appears that the rate of decrease is initially (up to the early 20 s) faster. The change in *k* with age is not simply an effect of decreasing cortical thickness but, rather, is an emergent effect from correlated changes in all three primary variables of cortical thickness, total surface area, and exposed surface area.

Altogether, cortical folding is a reminder that the organization of the brain and its connectome is constrained by the physical embedding in space. Mechanical factors limit and shape the wide variety of folding patterns that we see within and between species. Approaches from mathematics, physics, and computer science can be applied to measure, predict, and simulate the developmental changes of brain morphology. However, it is important to keep in mind that mechanisms that work at the local level, explaining local curvature and thickness within the gray matter, might not always be sufficient to explain the global level of cortical folding. Understanding local and global mechanisms is also relevant for clinical applications. As cortical folding and cortical thickness change for many brain disorders, a better understanding of the developmental factors might yield biomarkers for the early detection of abnormal development.

III Connectome Changes

The field of connectomics has expanded rapidly. Whereas a search in PubMed, which covers most but not all biomedical literature, yielded around one new article every week on brain connectivity in 2005, there are now more than 50 articles. Indeed, brain connectivity, especially functional connectivity, has replaced brain activity as main abstract keyword for the annual Human Brain Mapping conference. While functional and structural connectivity are strongly linked, we will mainly focus on structural connectivity changes in this part.

Following connectome maturation, the structure of the connectome is not static. Rather, it represents a fluid architecture that can be altered. There are changes as part of the normal aging process, but there are also changes due to neurodevelopmental and neurodegenerative disorders. However, not all changes are linked to cognitive deficits or disorders; some changes are compensatory or help to reduce disease symptoms. Learning is another example where structural and functional connectivity can change at all stages of the life span. Brain networks also show changes due to injury, such as strokes, lesions, or traumatic brain injury. Again, some changes help to retain or recover healthy brain function. Finally, brain stimulation can change the dynamics of network activity but, on the longer term, can also influence structural and functional connectivity.

In this part, we will look at concepts of connectome changes, examples of such changes, and computational models that can help to predict the effects of internal or external factors on brain networks.

11 Development and Aging

11.1 Overview

Human brain development is characterized by a protracted trajectory that extends into adulthood. Evidence from MRI has indicated a reduction in gray matter volume and thickness across large areas of the cortex and changes in subcortical structures. However, a recent study of ventral temporal cortex indicates that reduced thickness could be due to a blurred gray matter/white matter boundary, through reduced contrast in structural MRI, due to increased myelination (Natu et al., 2019). These findings suggest that cortex does not thin during childhood but instead gets more myelinated. This would match with white matter volume increases with age (Giedd et al., 1999) which could again reflect increased myelination of axonal connections (Sowell et al., 2004).

In addition to volume changes, connectivity changes of axonal fiber bundles have been investigated using DTI, observing measures of fiber integrity through estimates of fractional anisotropy and mean diffusivity, which presumably relate to changes in axonal diameter, density, and myelination (Jones, 2010). From childhood to adulthood, fractional anisotropy increases and mean diffusivity decreases in several major fiber tracts and brain regions (Tamnes et al., 2010; Lebel and Beaulieu, 2011).

The area of developmental connectomics spans from before birth through to early childhood (Cao et al., 2017). Emerging data on structural connectivity suggests that small-world, modular, and hub features in brain networks are already present during early development despite of appreciable anatomical changes during brain maturation. Structural and functional connectivity has been studied at very early stages, before birth, as part of the Developing Human Connectome Project (Fitzgibbon et al., 2016). Throughout the life span (Zuo et al., 2017), several studies in healthy populations are performed, including children and adolescents (cross-sectional Philadelphia Neurodevelopmental Cohort, ages 8–21 years; Child Mind Institute Healthy Brain Network, ages 5–21 years), adolescents and young adults (longitudinal IMAGEN Europe, ages 14–22 years), and a wider life span (Nathan Kline Institute/Rockland, ages 6–85 years). Questions regarding the role of genetic factors versus environmental factors for connectome development are being addressed within the twin study of the Human Connectome Project where young adult (ages 22–35 years) twins and nontwin controls are recruited (Van Essen et al., 2012). Focusing on aging, the UK Biobank Imaging project recruits 100,000 subjects ages 40 to 69 years, using MRI scans to assess structural and functional connectivity (Miller et al., 2016). We will briefly summarize some results from these studies and some developmental principles for healthy connectome maturation.

11.2 Network Changes during Early Brain Development

Pruning at the Local Neuronal Level

One of the crucial and universal characteristics of brain development is that initial exuberance of neurons, dendrites, axons, and synapses is followed by selective elimination. For example, synaptogenesis peaks in neonates and decreases afterward. Neuronal density is highest around birth and is, at that stage, about double that of an adult brain (Cowan et al., 1984). Synaptic density is at its highest around 1 to 2 years of age and decreases throughout maturation (Huttenlocher, 1979; LaMantia and Rakic, 1994). Gray matter volume and thickness peak around 10 years of age and decrease through adulthood. These phenomena are related to each other and especially pruning of synapses, dendrites, and axons is suspected to be the major underlying driving force for developmental anatomical changes in the brain.

Selective elimination of connections helps the nervous system to reorganize its network from a transient and redundant structure to an efficient and economic network, which provides neural plasticity for learning and memory and for repairing damaged circuits (Luo and O'Leary, 2005). Computational models have also demonstrated that brain networks become more efficient by selectively eliminating synapses (Chechik et al., 1999). Previously, it was believed that this pruning process only continues until puberty (Huttenlocher, 1979). Recent findings, however, demonstrated that synaptic pruning is prolonged into the third decade of life, especially in prefrontal areas (Petanjek et al., 2011).

Reduced Global Structural Connectivity between Brain Regions

Interestingly, large-scale brain connectivity studies have also observed "macroscopic pruning," which can be viewed as indirect evidence of pruning of synapses, axons, or dendrites influencing large-scale brain connectivity. There seems to be overconnectivity in the early developmental stage, and "pruning" of connectivity follows during development (Lim et al., 2015).

Interestingly, the changes before or around birth seem to be the opposite of those observed from childhood to adulthood (Tymofiyeva et al., 2013; van den Heuvel et al., 2015). Baby connectome studies have shown that modularity and clustering coefficient increased from preterm to term babies (Tymofiyeva et al., 2013; van den Heuvel et al., 2015), whereas modularity and clustering coefficient decrease during adolescence and adulthood, leading to a more distributed and integrated network (Hagmann et al., 2010), which may result from eliminating diffusive and less accurate connections.

Early Rise of Hubs and Modules
The human brain shows rapid growth during the prenatal period and the first few postnatal years. Development of brain networks follows a sequential order from primary to higher-order systems and from a tendency of network segregation to network integration (Zhao, Xu, and He, 2019).

Observation of structural connectivity changes between the ages of 2 and 18 years showed an increase in node strength and efficiency along with a decrease in clustering coefficient with age (Hagmann et al., 2010). Major hubs and modules were already apparent at the age of 2 years and further strengthened afterward. Finally, structural connectivity and functional connectivity were strongly correlated. This correlation further increased with age, potentially because of changes in axonal diameter and myelin thickness. Overall, structural networks show an improved ability for information integration through hubs at the global and modules at the local level. The extent to which these patterns also occur for functional connectivity, however, is controversial (Betzel et al., 2014; Váša et al., 2019).

Postnatal Changes in Cortical Thickness, White Matter Volume, and Cortical Folding
The human cortical gray matter surface area increases threefold between term birth and adulthood. By term gestation, almost all neurogenesis and neuronal migration are complete. Still, postnatal surface expansion differs between brain regions (Hill et al., 2010). Regions that show high expansion are functionally and structurally less mature, compared to low-expansion regions (see figure 11.1). The more mature low-expanding regions have more simply branched dendrites and fewer spines, indicating that major local refinements already took place. Looking at white matter expansion, subcortical and occipital regions expand at higher rates than frontal, anterior temporal, and parietal that show a more protracted maturation.

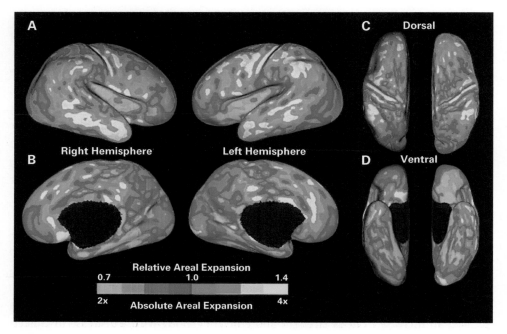

Figure 11.1
Postnatal cortical surface expansion. Maps of postnatal cortical surface expansion on the standard mesh average inflated term infant surfaces for both hemispheres, shown in lateral (A), medial (B), dorsal (C), and ventral (D) views. The absolute expansion scale indicates how many times larger the surface area of a given region is in adulthood relative to that region's area at term. The relative expansion scale indicates the difference in proportion of total surface area at term birth and adulthood (after Hill et al., 2010).

Along with white matter volume and myelination, cortical folding is also changing. Gyrification of the cortex peaks around 5 or 6 postnatal months and gradually decreases throughout adulthood (Armstrong et al., 1995). When ages from 12 to 23 years are looked at, GI and cortical thickness vary during adolescence (Klein et al., 2014). GI values reduce with age in several cortical regions including precentral, temporal, and frontal areas. These reductions in local cortical folding could not be fully explained through changes in gray matter thickness, volume, or surface area, indicating that other mechanisms, possibly involving brain connectivity changes, must play a role.

Learning-Related Changes

In addition to ongoing changes as part of brain maturation, there are connectome changes due to learning. These changes are usually discussed in the framework of brain or synaptic plasticity. There are two ways in which connections can change over time. For structural plasticity, discussed in detail in chapter 14, connections appear or disappear over time. For functional plasticity, the weight of an existing connection changes but synapses are not removed. More detailed descriptions of learning mechanisms, including mathematical and computational aspects, can be found in, for example, Dayan and Abbott (2001), Koch (2004), Trappenberg (2010), and Sterratt et al. (2011).

A prime example for learning relates to the language abilities of humans. While some language abilities are innate, there is a prolonged learning effort to become able to read, write, and speak. This effort involves visual, auditory, somatosensory-motor, and higher cognitive areas. For literacy, building up the link between written symbols and spoken language, the left ventral occipitotemporal cortex and the left superior temporal regions are reorganized. These changes are linked to increased structural connectivity of the posterior corpus callosum and the left arcuate fasciculus (Dehaene et al., 2015).

When structural connectivity in illiterate individuals and individuals who obtained literacy during adulthood was looked at, literacy was related to an increase in fractional anisotropy and a decrease in perpendicular diffusivity in the temporoparietal portion of the left arcuate fasciculus (Thiebaut de Schotten et al., 2014). The extent of these changes correlated with reading performance. Therefore, reinforcement for left temporoparietal connections is crucial for gaining literacy.

Sleep is thought to be important to consolidate what has been learned during the awake phase. The formation and retention of associative memory has been studied in computational models that include both functional and structural plasticity. During sleep or rest, cell assemblies reactivate spontaneously, reinforcing memories against ongoing synapse removal and replacement. While the connectivity related to memory representations is strengthened, other connections decay or vanish. Resulting improvements in computational models are consistent with memory gains observed experimentally (Fauth and van Rossum, 2019).

It is important to note that sleep is also enhancing brain function through the removal of waste material in addition to removing "waste" connectivity. During sleep or anesthesia, the exchange of cerebrospinal fluid with the interstitial fluid between neurons is increased, leading to improved removal of β-amyloid and other interstitial proteins (Xie et al., 2013). This clearance of potentially neurotoxic waste products that accumulate in the awake stage could be another component of improving learning through sleep or rest.

Altogether, this indicates that sleep disruption might lead to severe cognitive deficits. Indeed, β-amyloid pathology is associated with both non–rapid eye movement

(NREM) sleep slow-wave activity and memory impairment in older adults. Sleep disruption might therefore be a mechanistic pathway through which β-amyloid pathology may contribute to hippocampus-dependent cognitive decline in the elderly (Mander et al., 2015).

Notably, the strength of synapses can also change spontaneously: the ability of a synapse to retain its strength over time, given no change in input, is called synaptic tenacity (Ziv and Brenner, 2018). Observing synapses over many days, in spontaneously active networks, distributions of synaptic sizes were generally stable. However, individual synapses were continuously and extensively changing their size. For active networks, particularly for synchronous activity periods, large synapses grew smaller whereas small synapses grew larger. This resulted in synaptic changes affecting the size of their excitatory input, but these changes, albeit at slower rate, were also observed after activity was blocked. Altogether, these changes can help to limit the activity of neurons and therefore be a potential mechanism for synaptic homeostasis (Minerbi et al., 2009).

11.3 Maintaining Network Features that Enable Cognition

Faster Maturation of Hubs during Early Development

Using structural connectivity of preterm and full-term neonates, scanned at 31.9 to 41.7 postmenstrual weeks, the fastest increases in node efficiency mostly appeared for network hubs in primary sensorimotor regions, superior middle frontal regions, and precuneus regions (Zhao, Mishra, et al., 2019). At the same time, edge strength particularly increased within modules, for short-distance connections, and for nodes belonging to the rich-club network of highly connected nodes. Overall, this indicates that connections that are crucial for information integration, either at the local level of modules or at the global level of the rich-club network, are strengthened during early brain development. This early hub formation is also in line with the old-gets-richer model described in chapter 8 and the potential early formation of network hubs in *C. elegans* (Varier and Kaiser, 2011).

Preferential Detachment: Retaining Long-Distance, Intermodular, and Interhemispheric Links

A study of structural connectivity changes between the ages of 4 and 40 years shows that the reduction in connection strength, measured as number of streamlines in deterministic tracking, is not random but highly preferential (see figure 11.2): fewer long-distance, thin, and intermodular fiber tracts showed streamline loss than would be

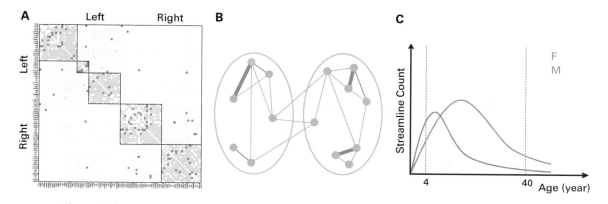

Figure 11.2
Preferential loss of structural connectivity (after Lim et al., 2015). (A) Changes in fiber tract strengths (gray: intramodular edges; light gray: intermodular edges, both without changes over age; red: edges with a decreased streamline count; blue: edges with an increased streamline count; and yellow: edges with sex-specific changes). (B) Types of fiber tracts that are reduced (schematic with ellipses representing both hemispheres): red lines are where the reduction of streamlines occurred with preferential loss for thick, short-distance, or intramodular tracts. (C) Hypothetical developmental curves for males (blue) and females (red): for the total streamline count, there is a longer lasting and higher peaked increase and a delayed decrease in males.

expected given how often such fiber tracts could have been affected by chance (Lim et al., 2015). Moreover, this refinement of structural connectivity occurred earlier for girls than for boys.

This preferential streamline loss has several implications for the stable topological features. First, small-world features were retained over age despite the overall reduction in the number of streamlines. A significant decrease in many long-distance streamlines would remove shortcuts and result in larger path lengths and reduced global efficiency while fewer connections between neighbors would decrease local clustering and local efficiency, disrupting small-world features of a brain network. However, global efficiency stayed comparable to that of rewired networks while local efficiency was much higher than in rewired networks across age, conserving small-world topology (Latora and Marchiori, 2001).

We would therefore expect changes mainly in short-distance connectivity. Indeed, short streamlines were mostly affected, and long-distance connectivity was rather preserved. Relatively conserved streamlines in long-distance fiber tracts could be achieved by strengthening long-range pathways in the brain network and a reduced number of streamlines in short fiber tracts could be due to weakening of short connections.

Moreover, in line with previous rs-fMRI and DTI studies (Fair et al., 2009; Hagmann et al., 2010), modularity Q remained stable over age. This retained modular organization (Kaiser and Hilgetag, 2010; Meunier et al., 2010) might be crucial in keeping the balance between information integration and the segregation of separate processing streams (Sporns, 2011). Too many connections between modules would interfere with different processing demands—for example, leading to interference between visual and auditory processing. In addition, more intermodule connections would also facilitate activity spreading, potentially leading to large-scale activation as observed during epileptic seizures (Kaiser, Görner, and Hilgetag, 2007). However, because of the reduction of streamlines in intramodule edges, proportionally intermodule connections increased, indicating that brain networks became more distributed with age as observed in previous studies (Fair et al., 2009; Supekar et al., 2009; Hagmann et al., 2010).

Sex-Specific Differences in Structural Connectome Maturation

A study of structural connectivity in 949 subjects with ages between 8 and 22 years identified distinct differences between males and females both in terms of connectivity and in terms of the trajectory of changes during development (Ingalhalikar et al., 2014). These trajectories already diverge in the youngest observed age group, the 8- to 13-year-olds, with further increased differences in the older age groups. In general, males had greater within-hemispheric connectivity, as well as enhanced modularity and transitivity, whereas between-hemispheric connectivity and cross-module participation predominated in females. This suggests that male brains have increased connectivity between perception and coordinated action regions whereas female brains have increased connectivity between analytical and intuitive processing regions. Finally, regions in the frontal, parietal, and temporal lobes had higher participation coefficients—a higher proportion of their connections linking to other lobes—in females than in males. The cerebellum was the only region that displayed higher participation coefficients in males.

Features Related to Intelligence

Certain cognitive and behavioral performance measures in later life can already be predicted from topological network features during early development (Zhao, Xu, and He, 2019). There are many factors that can influence network topology, ranging from sex (see above) to socioeconomic situation to education. While there are many cognitive performances that could be analyzed, we will here discuss intelligence in more depth as it is the focus of many studies in the field. One measure of intelligence is the intelligence quotient (IQ), consisting of scores for fluid intelligence, the ability to solve novel problems, and crystallized intelligence, the use of past experience to solve problems.

Studies of brain activity have indicated that fluid intelligence involves the frontal cortex (Isingrini and Vazou, 1997; Duncan et al., 2017). The parieto-frontal integration theory describes a global network of distributed brain regions as influencing intellectual ability (Jung and Haier, 2007; Tschentscher et al., 2017).

Alongside functional signatures of intelligence, studies have observed differences in gray matter volume across IQ, with higher IQ associated with overall increased cortical volume (Haier et al., 2004; Sabuncu et al., 2016).

Better Network Integration for Higher Cognitive Performance

Structural connectivity organization, involving shorter path lengths and network hubs, is indicative of academic attainment in reading and mathematics (Bathelt et al., 2018). For rs-fMRI functional connectivity, shorter local path lengths, measuring the length of the shortest paths from one region to all other regions in the network, were for specific regions also predictors of IQ (van den Heuvel et al., 2009). Furthermore, for structural covariance networks, gifted children showed more integrated networks where nodes had higher participation coefficients—a higher proportion of their connections linking to other modules than to its own module—indicating greater intermodular communication mediated by connector hubs with links to many modules (Sole-Casals et al., 2019). Moreover, for gifted children, more connector hubs were found within the association cortex.

Overall, this indicates that better information integration and/or information broadcasting is crucial for better cognitive performance. However, there are several pieces still missing to elucidate the link between network structure and the development of cognitive performance. First, unless we have postmortem information, we do not know the direction of fibers, so we cannot decide whether hubs act more as integrators or more as broadcasters. Second, synchronous activity is crucial to link processing in different regions—for example, forming and maintaining phase relations between regions, a theory of how brain regions interact known as communication through coherence (Fries, 2015). Without information about conduction delays across fiber tracts, relating to both fiber trajectory length and myelination, it is difficult to assess how connectivity changes relate to synchronous processing changes. Finally, the relation of a network feature with cognition does not inform us about the detailed role of that feature within information processing in the brain. Longitudinal studies observing how network features coevolve with cognitive abilities, experimental studies of functional connectivity during task performance, and computational models of developing brain function will be needed to address this question.

12 Neurodevelopmental Disorders

12.1 Overview

During normal brain development, long-distance connections, hubs, and small-world and modular features are preserved to yield a balance of integrated and separated information processing. Consequently, any changes in these crucial features might be linked to the cognitive deficits that we observe for developmental brain disorders. There is a growing literature on changes in both global features and individual connections in brain disorders, with *pathoconnectomics* (Rubinov and Bullmore, 2013) being one term to describe the network neuroscience of these conditions. Within this chapter, we look at neurodevelopmental conditions where behavioral symptoms arise within the first 40 years of life.

Studies in both structural and functional connectivity reported aberrant connectivity possibly caused by imperfect maturation of a brain network in neurodevelopmental disorders such as schizophrenia, the autistic spectrum, and epilepsy. It is important to note that altered features could either be linked to cognitive deficits observed in these diseases or, in contrast, be beneficial alterations that compensate for disruptive changes elsewhere in the network.

Recent efforts to understand brain network changes have focused on deviations of global network features from healthy controls. For functional connectivity, a deviation from a small-world architecture, moving either toward a more random or a more regular (lattice) organization, was proposed to be linked to cognitive and psychiatric disturbances (Reijneveld et al., 2007). Related to global information flow, it was proposed that reductions in brain connectivity can lead to more information being routed through hubs. In turn, this could lead to an "overload" and, as a consequence, a failure of hub nodes leading to the observed deficits in brain function (Stam, 2014).

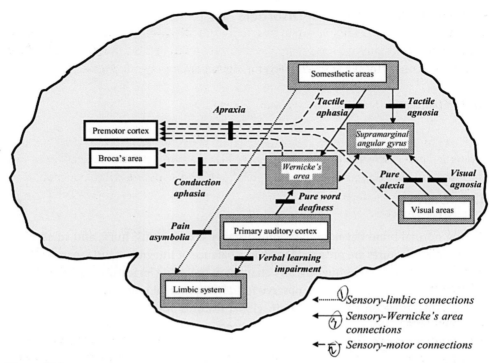

Figure 12.1

Geschwind's disconnection syndromes. The pathways implicated in the principal syndromes described by Geschwind, classified into three types: sensory–limbic disconnection syndromes (dotted lines), sensory–motor disconnection syndromes (dashed lines), and sensory–Wernicke's area disconnection syndromes (solid lines) (after Catani and ffytche, 2005).

Another approach is to highlight the role of individual connections or regions in disease symptoms. At the extreme case, one might observe what happens if entire network components are removed. Such lesion-deficit models are linked to the ideas of disconnection (originally defined as "disconnexion" by Geschwind, 1965): a disconnection lesion will be a large lesion either within a region or of the white matter projecting from that region. These disconnections were originally discussed in relation to neurological conditions such as language deficits (tactile aphasia, pure word deafness, pure alexia, or agnosia), learning and emotional response, and sensorimotor integration (see figure 12.1). However, disconnection has also been discussed for developmental diseases such as schizophrenia (Friston, 1998). Disconnection, in the literal sense of reduced connectivity between nodes, might be a too narrow interpretation of changes for brain disorders. Therefore, the term "dysconnection," encompassing both increases and decreases in connectivity, would be a more suitable term to describe the observed malformation of brain networks (Stephan et al., 2006). In the following, we will look at malformed connectivity for several developmental brain network diseases.

12.2 Changes for Neurodevelopmental Disorders

Changes for network disorders can involve any single or combination of topological or spatial features discussed above. However, we here want to focus on three changes that were already studied for developmental disorders and where network concepts have been introduced in the previous chapters: changes in hub organization, changes in modular architecture, and changes in long-range connectivity. There is a large overlap in network changes between different disorders, and not all changes, identified when comparing patient and control groups, can predict a disorder at the individual level. In this fast-moving field, we can only show a small selection of studies and developmental brain disorders.

Schizophrenia

Schizophrenia is a psychiatric disorder with first clinical symptoms occurring during the teen years or early adulthood. Changes in brain function in patients are assessed with the Positive and Negative Syndrome Scale (PANSS) with positive features being the ones that are present in patients but not in healthy controls—for example, hallucinations—and negative features being the ones that are lacking in patients—for example, reduced ability to express emotions.

Schizophrenia has been thought of as a neurodevelopmental disease caused by disconnectivity (Boksa, 2012) especially in the frontal and temporal lobes (Pettersson-Yeo et al., 2011). A wide range of structural connectivity changes within and between regions includes aberrantly located neurons in the white matter, implying disturbances in neuronal migration and formations of connections; abnormal cytoarchitecture of the entorhinal cortex, implying aberrant formation of microcircuits; white matter abnormalities observed with diffusion weighted imaging; and changes in corpus callosum morphology (see overview in Stephan et al., 2006).

Connectome studies have found changes in small-world and modular organization (Micheloyannis et al., 2006; Liu et al., 2008; Bassett et al., 2008), and other functional connectivity measures (Woodward et al., 2011). Changes include both increased and decreased connectivity (Fornito et al., 2012); however, decreased connectivity has been reported consistently across all stages of schizophrenia and in various modalities

(Pettersson-Yeo et al., 2011); child-onset schizophrenia patients showed fewer short-range functional connections than healthy controls (Alexander-Bloch et al., 2013), leading to decreased modularity and clustering coefficients in patients (Alexander-Bloch et al., 2010). When structural connectivity was used to assess connection fingerprints of different regions, the inferior frontal gyrus and the putamen mainly contained relatively increased connection probabilities to areas in the frontal, limbic, and subcortical areas while the medial frontal gyrus showed only reduced connectivity (Thanarajah et al., 2016).

In a longitudinal DTI study over a 5-year period, a significant deficit of global integration was found in patients, which further increased over time (Sun et al., 2016). Moreover, several brain regions (e.g., the inferior frontal gyrus and the bilateral insula) that are crucial for cognitive and emotional integration were aberrant. The increasing characteristic path length or reducing global efficiency over time in patients was also associated with worsening clinical symptoms as measured by the PANSS scores.

Genetic factors play a huge role: children where one parent has schizophrenia are 10 times as likely to develop the disease than are children where none of the parents suffered from schizophrenia (Collin et al., 2017). There are lower levels of anatomical rich-club connectivity in nonpsychotic young offspring of schizophrenia patients. This finding suggests that the brain's anatomical rich-club system is affected in at-risk youths, reflecting a connectome signature of familial risk for psychotic illness.

Autism Spectrum Disorder
Autism shows early social, physical, and language symptoms within the first 3 years of life and is 4 times more common in males. The autism spectrum includes autism, Asperger's syndrome, and pervasive developmental disorder. Epilepsy is common in children with autism.

Concerning morphological changes, people with autism spectrum disorder show accelerated childhood cortical expansion, accelerated cortical thinning in late childhood/adolescence, and slowed-down cortical thinning in early adulthood (Zielinski et al., 2014). At the microcircuit level, there are visible abnormalities in frontal lobe organization (Courchesne and Pierce, 2005). Indeed, disrupted communication between the frontal lobe and other parts of the brain is a hypothesis for how functional deficits in this disorder can arise.

For structural connectivity, reduced local efficiencies in low-level sensory processing regions have been observed in high-risk infants as young as 6 months of age and correlate significantly with 24-month symptom severity (Lewis et al., 2017). Later in development, patients with autism spectrum disorder fail to show the increases in rich-club organization observed in typically developing subjects (Watanabe and Rees, 2015). Even for high-functioning adolescents with autism spectrum disorder, there are still local connection strength changes in the cingulate cortex (Ball et al., 2017).

Functional connectivity shows decreased long-distance connectivity between frontal and posterior regions as well as between hemispheres and increased short-distance connectivity (Belmonte et al., 2004; Williams and Minshew, 2007; Anagnostou and Taylor, 2011; Paul, 2011; Lau et al., 2013). When the hierarchical organization of resting-state functional connectivity was looked at, there were also a reduced ability to activate the rich-club network and disruptions of long-range connections (Hong et al., 2019).

A longitudinal study, observing structural connectivity in young patients and controls at baseline and 3 to 7 years later, shows reduced connectivity within the frontoparietal network—and its broader connectivity—in autism spectrum disorder during adolescence and early adulthood (Lin, Perry, et al., 2019). During the same period, these connections strengthened in controls. Finally, the extent of frontoparietal changes in patients was linked with the autism phenotype development over time, indicating that these connectome features could be a potential predictor of autism progression.

Tourette's Syndrome

Gilles de la Tourette syndrome is characterized by tics, first occurring during childhood. While these symptoms can persevere into adulthood, a remission of symptoms can be observed in many patients. Structural connectivity in adult patients showed reduced connectivity in right-hemispheric networks and reduced local clustering coefficient, efficiency, and strength, but increased normalized global efficiency (Schlemm et al., 2017). This increased global efficiency was related to the severity of tics. Moreover, for structural connectivity in children with Tourette's syndrome, there is an increased rich-club backbone, potentially explaining the higher global efficiency in child and adult patients (Wen et al., 2017).

Some of the observed changes might be due to compensatory alterations. In a behavioral study, patients showed increased control of motor output with structural connectivity changes in the corpus callosum and forceps minor being linked to the severity of tics (Jackson et al., 2011). These compensatory changes leading to enhanced self-regulation are supposed to be due to local increases in "tonic" inhibition leading to a reduction in the "gain" of motor excitability (Jackson et al., 2015).

Epilepsy

There are many different subtypes of epilepsy, but common characteristics include the occurrence of spontaneous recurrent seizures, often with a loss of consciousness, associated with abnormal brain activity. First episodes of abnormal brain activity can occur at any age, from neonates and infants to children and adolescents. A particular focus group for brain network studies in patients are the ones where antiepileptic drugs do not work (medically intractable) and where the surgical removal of brain tissue is one of the remaining options.

Epilepsy affects the modular organization of functional connectivity (Chavez et al., 2010; Vaessen et al., 2013) and the link between structural and functional connectivity (Zhang et al., 2011); it is also associated with decreased long-distance connectivity and increased structural short-distance connectivity (Bonilha et al., 2012; DeSalvo et al., 2014). Furthermore, idiopathic epilepsy patients suffer from underdeveloped white matter (Hermann et al., 2006; Hutchinson et al., 2010) impeding long-range connectivity development. However, one needs to keep in mind that many of the structural connectivity changes can be explained by a decreased surface area and increased streamline count in patients, as shown for temporal lobe epilepsy (Taylor, Han, et al., 2015).

The observed connectome changes might be either causes or consequences of epileptic seizures. In general, developmental changes such as focal cortical dysplasia, lissencephaly, heterotopia, and polymicrogyria can lead to hyperexcitability of tissue and epileptic seizures later on. An imbalance between excitation and inhibition can result from defects in cell proliferation in the germinal zone, impaired neuronal migration and differentiation, and delayed or reduced arrival of inhibitory interneurons into the cortical plate (Bozzi et al., 2012). Moreover, defects in later elimination and remodeling of synapses during early critical periods can also lead to hyperexcitability.

Depression
Major depressive disorder is characterized by mood changes leading to ruminative, slow, and monotonous thinking. In addition to developmental origins, recent studies have emphasized the role of inflammation on changes in brain connectivity (Schrepf et al., 2018), and neuroinflammation is discussed as one possible cause for depression (Bullmore, 2018).

An analysis of dynamic functional connectivity reveals that the depressed brain shows an abnormally stable, synchronous pattern of activity (Demirtas et al., 2016). Moreover, there is increased functional connectivity in the default mode network. On the other hand, functional intrahemispheric (left and right) and interhemispheric (heterotopic) connectivity is reduced, leading to decreased local, global, and normalized global efficiency for both hemispheric networks; increased normalized local efficiency for the left-hemispheric networks; and decreased intrahemispheric integration and interhemispheric communication in the dorsolateral superior frontal gyrus, anterior cingulate gyrus, and hippocampus (Jiang et al., 2019). Finally, the functional–structural coupling of intrahemispheric connections is decreased, and the extent of this decrease correlates with disease severity.

12.3 Causes of Developmental Disorders

Understanding the developmental causes of connectome changes in brain disorders is still at an early stage. While the previous section has described how pathological connectomes differ, there are often multiple potential pathways through which such changes could arise during brain development. Longitudinal studies, before and after the onset of clinical symptoms, will be crucial when only the time course of changes can distinguish between different underlying mechanisms. In this section, we will discuss potential causes of developmental disorders based on computational and experimental studies.

Altered Modular Organization

Several developmental brain disorders show changes in their modular architecture with reduced connections between modules and, in some cases, increased connectivity within modules. One way to observe such changes is to look at the modularity Q (see chapter 2). Another measure is the dispersion D, a measure to assess the degree to which a node connects to nodes that belong to another group of nodes, nodes in a different region, or nodes in a different module. Dispersion does not simply measure the relative number of connections that link to other modules (the measure for this would be the participation coefficient; see Sporns and Betzel, 2016), but to what proportion of other modules the node is connected. If there is a high degree of targeted wiring, nodes within one region of the network should connect to few other regions. That means, instead of randomly following a direction, they should follow existing fibers that lead to an already connected region. For the human connectome, individual nodes could be regions of interest (ROIs), and regions could be cortical or subcortical areas. We then define dispersion as the fraction of regions (brain areas) with which a node (ROI) is connected (Kim and Kaiser, 2014): the dispersion is low if an ROI is only connected to ROIs within a few regions; the dispersion is 1 if the ROI is connected to other ROIs within its own region as well as with ROIs in all other regions (see figure 12.2).

For human structural connectivity in healthy subjects with 998 ROIs (nodes) and 66 brain areas (regions), the dispersion is 0.12, which means that an ROI within a brain area is, on average, only connected to 12% of all brain areas (i.e., itself and seven other

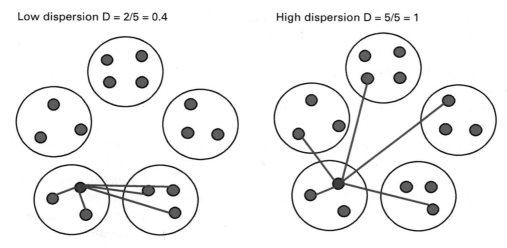

Figure 12.2
Dispersion (D) as a measure of reaching multiple regions or modules. Left: A node is linked to two out of five network regions: other nodes in the same region where the node is situated and nodes in a different region. The dispersion of connections from that node is 40%. Right: The same node is linked to all five regions of the network leading to a dispersion of 100% (cf. Kim and Kaiser, 2014).

areas). For a rewired network with the same modularity Q, the dispersion is much higher, around 30% (Kim and Kaiser, 2014), leading to a more diffuse pattern of connections. The reduced dispersion in the original, nonrewired network leads to a higher characteristic path length, potentially reducing interference between different processing pathways and different sensory modalities.

Given the relationship between dispersion and other network properties that change in schizophrenia (Zalesky et al., 2011), autism (Courchesne and Pierce, 2005), or epilepsy (Lemieux et al., 2011), a reduced coherence of fiber tracts might be an important component in the path toward developmental diseases. Moreover, the dispersion might be related to changes in diffusion weighted imaging, since a more distributed pattern of connectivity would break apart the fascicular pattern of fiber tracts. Therefore, we would expect that higher values of dispersion will be associated with lower values of fractional anisotropy and with a shift toward networks with lower characteristic path length. For neural disorders, for example, lowered fractional anisotropy was reported for partial intractable epilepsy (Dumas de la Roque et al., 2005), autism spectrum disorders (Sundaram et al., 2008), and schizophrenia (Skudlarski et al., 2010; van den Heuvel et al., 2010). Note, however, that lower fractional anisotropy might result not only from more diffuse fiber tracts within a voxel but also from reduced myelination (Jbabdi and Johansen-Berg, 2011; Jbabdi et al., 2015).

Pruning-Related Changes

Various neurodevelopmental disorders have been related to abnormal pruning of synapses. Failure to prune away weak synapses was shown to impede constructing synaptic multiplicity, where one neuron established multiple synapses with another neuron. This in turn leads to weakening of long-range functional connectivity and is correlated with autism-associated behavior in knockout mice (Zhan et al., 2014).

Failure of pruning excitatory synapses during postnatal development causes hyperexcitability (Galvan et al., 2000; Caleo, 2009; Zhou et al., 2009; Bozzi et al., 2012). This could be related to diseases that show an imbalance between excitatory and inhibitory synapses such as autism (Rubenstein and Merzenich, 2003), social dysfunction (Yizhar et al., 2011), and epilepsy (Stief et al., 2007).

At the scale of the macroconnectome, changes in long-range connectivity, due to pruning or other mechanisms, can also be found in epilepsy. For idiopathic generalized epilepsy, seizure-like dynamics were reproduced in a computational brain network model with enhanced corticoreticular connections along with impaired corticocortical connections (Sinha et al., 2019). Indeed, such structural connectivity changes were observed in patients suffering from this disease.

Plasticity-Related Changes: The Disconnection Hypothesis for Schizophrenia

Changes in networks could result not only from structural plasticity, the removal of synapses or axons, but also from functional plasticity, where synaptic weights are adjusted. One example is the idea of the disconnection hypothesis in schizophrenia as an explanation for cognitive deficits in functional integration (Friston, 1998).

The disconnection hypothesis assumes that activity-dependent plasticity is disturbed at later stages of development closer to the onset times of symptoms in adulthood. Therefore, parts of the brain that mature earlier, such as the visual cortex, develop normally whereas other regions that mature later, such as the frontal lobe, are affected. Indeed, structures in the visual system such as ocular dominance columns appear normal in patients with schizophrenia. Another prediction, if plasticity is the cause of connectome changes, is an important role for neurotransmitters. Indeed, there is decreased hippocampal expression of the embryonic form of the neural cell adhesion molecule in schizophrenia (for an overview of supporting evidence, see Stephan et al., 2006; Pettersson-Yeo et al., 2011).

Malnutrition

There is also an environmental influence on connectome development. Deficits in nutrition, both before and after birth, can affect brain development and behavior. Reduced protein intake of the mother during pregnancy can lead, in the offspring, to increase in anxiety; problems in social play, motivation, and exploratory activity; and disruptions of sleep (Barbeito-Andres et al., 2018).

Concerning the structural connectome following malnutrition, the rich-club architecture is reduced, and long-distance connections are preserved while short-distance connections are reduced (Barbeito-Andres et al., 2018). The growth of temporoparietal cortex is relatively preserved, and also several network features, such as degree, clustering coefficient, and characteristic path length, remain mainly unchanged.

Delayed Development

Onset times of developmental diseases, ranging from autism spectrum disorders and anxiety disorders to schizophrenia and bipolar depression, range from the first year after birth to (early) adulthood. Indeed, delays in connectome maturation were recently linked to psychiatric diseases (Kaufmann et al., 2017). Cortical thickness in ADHD patients, for example, peaks at a later age than for healthy subjects. How does timing influence different subtypes of a developmental disease such as schizophrenia?

Early onset schizophrenia (EOS), before the age of 18 years, is characterized by changes in structural and functional hubs, in the default mode network, and by more gray matter in the frontotemporal network (Yang et al., 2014). At the same time, long-range callosal fibers between hemispheres are reduced, and homotopic regions in both hemispheres show more divergent connectivity patterns (Li et al., 2015). Finally, compared to adult onset schizophrenia, EOS shows reduced local and remote hub connectivity (Jiang et al., 2015).

In light of the mechanisms that we discussed for the formation of connectome features, a delay in the development of nodes within a hemisphere could explain the reduced connectivity in EOS. Slower nonlinear network growth, where fewer nodes form at each step, will reduce the connectivity of network hubs and could explain the reduced connectivity in the default mode network and of structural and functional hubs. Finally, a later start of maturation could also lead to a longer duration of maturation. In that way, longer-lasting tissue growth might partially counteract the effect of pruning and lead to an increased amount of gray matter.

A delay is only one possibility for how the time course of normal and pathological brain development could differ. The timing of events in brain disorders might be too fast (precocious), be halted (stopping further development), fail to reach the level of mature development, or be ectopic for changes that are absent during normal development (Di Martino et al., 2014).

Overall, the underlying biological changes in neurodevelopmental disorders affect multiple brain regions, leading to the idea of brain network disorders. Concepts from graph theory such as rich-club, small-world, modular, and spatial features can be used to characterize and classify disorders. Besides diagnosis, network neuroscience can also be applied to inform treatment, identifying targets for surgery in epilepsy (Goodfellow et al., 2016; Sinha et al., 2017) or for brain stimulation in depression or obsessive-compulsive disorder (see chapter 15). Novel approaches are combining network analysis with network control theory to assess the impact of interventions (Bassett et al., 2018). However, determining the developmental origins of the observed connectome changes, along with generative and growth models, could add an important component for better diagnosis and intervention planning in the future.

13 Neurodegenerative Disorders

13.1 Overview

Dementia is a term used to describe a group of symptoms related to cognitive and functional decline, usually with a neurodegenerative etiology, that affects almost 50 million people worldwide (Geser et al., 2005; American Psychiatric Association, 2013). The most common cause of dementia in older adults is Alzheimer's disease (AD) dementia, with 50% to 70% of clinically diagnosed cases (McKeith et al., 2007), followed by dementia with Lewy bodies (DLB), which accounts for 4% to 8% of the cases (McKeith et al., 2007), and Parkinson's disease dementia (PDD), which develops in ≈80% of people with Parkinson's disease longitudinally (Hely et al., 2008). The main initial symptom of AD consists of episodic memory loss which occurs gradually over at least a 6-month time frame prior to seeing a clinician (Grober and Buschke, 1987; Dubois et al., 2014). Similarly to AD, cognitive impairments, more related to executive control than memory deficits, develop at early stages in DLB, and it can be difficult at this stage to separate the conditions (Palmqvist et al., 2009). Detecting the early core clinical features of DLB, which include cognitive fluctuations, parkinsonism, visual hallucinations, and REM sleep behavior disorder (McKeith et al., 2017), may increase the accuracy of the clinical diagnosis. At the onset of cognitive impairment or at later stages (McKeith et al., 2005), DLB patients often develop Parkinsonian symptoms such as bradykinesia, tremor, and rigidity (Hornykiewicz and Kish, 1987; Gaig and Tolosa, 2009), a symptomatic spectrum similar to PDD. The common etiology of DLB and PDD is the progressive accumulation of intracellular alpha-synuclein across the brain, known as Lewy bodies.

Understanding the time course of these diseases using experimental and clinical studies as well as, more recently, computational models can give insights into the underlying disease mechanisms that lead to the observed connectome and behavioral changes. Looking at disease progression is of key interest for two reasons. First, one might use models of the progression from a healthy condition to dementia to find biomarkers that can detect the later onset of dementia before any clinical symptoms become apparent. Even though there are no direct medical interventions available, lifestyle changes such as increased physical and social activity might delay the onset of the disease (Fratiglioni et al., 2004; Lövdén et al., 2013). In addition, a better understanding of the underlying mechanism might inform the search for new treatments and the later choice of treatment once the clinical condition sets in. Second, models of disease progression after the onset of symptoms can indicate how fast cognitive function is expected to deteriorate. Again, such a predicted future trajectory for an individual patient can inform the choice of interventions.

13.2 Changes during Disease Progression

Histological Changes

Many neurodegenerative diseases are now thought to arise from the production and spreading of toxic proteins. The basis of this has arisen from work in the prion field. Prions ("proteinaceous infectious particles") are unconventional infectious agents consisting of misfolded prion protein (PrP) molecules. These molecules aggregate with one another and incite a chain reaction of PrP misfolding (for more information about molecular changes, see Jucker and Walker, 2013). Prions can split up and spread through the nervous system, leading to functional changes and, ultimately, cell death. Cell death leads to reductions in gray matter volume and white matter connectivity. Prion formation can arise because of genetic factors, infections, or spontaneous protein misfolding.

Different proteins are involved for each type of dementia and can be detected through histology in postmortem brain tissue of patients (Jucker and Walker, 2013): Amyloid-β deposits and tau inclusions are the predominant neuropathological motif for AD, Lewy bodies composed of intraneuronal cytoplasmic inclusions comprising aggregates of α-synuclein (a presynaptic protein) and ubiquitin (a heat-shock protein associated with protein degradation) for DLB, α-synuclein and Lewy bodies restricted to the substantia nigra and brain stem for Parkinson's disease, and TDP-43 (TAR DNA-binding protein) inclusions for frontotemporal dementia.

Sequence of Spreading throughout Brain Networks

Interestingly, the starting nodes within the brain and the sequence of how changes spread throughout the network often follow a distinct pattern for each disease (see figure 13.1). Braak et al. (2003) proposed a staging for progressive α-synuclein accumulation for Parkinson's disease, with Lewy body pathology initiating in the dorsal motor nucleus of the glossopharyngeal and vagal nerves and the anterior olfactory nucleus, spreading rostrally to the pons and midbrain, followed by the limbic system, and finally to the neocortex. This caudo-rostral Lewy body accumulation in PDD patients is not always the case in DLB patients, and thus the extent to which the Braak staging hypothesis is applicable to DLB is yet to be elucidated. The spreading throughout brain networks is a relatively slow process, and for AD it is presumed that pathological changes start to emerge 10 to 20 years before the onset of clinical symptoms. For this reason, both proteinaceous seeds in bodily fluids and changes in brain connectivity could be early biomarkers of later onset of dementia.

Why Do Symptoms Occur So Late?

Changes for AD start to occur more than 10 years before cognitive deficits in brain function arise. For Parkinson's disease, more than two thirds of dopaminergic cells in the substantia nigra are dead by the time motor deficits become apparent. The phenomenon that a system still maintains a high level of performance despite failure of many of its components is known as "graceful degradation" in computer science. For neural networks, this means that function can be retained (Belfore et al., 1989) or regained (Hinton and Sejnowski, 1986). Altogether, this indicates that changes in connectome structure, before any change in brain function, could yield an early indication of a later onset of dementia.

Changes in Structural and Functional Connectivity

Both reductions and increases of connectivity have been reported in dementia, and results critically depend on the modality (EEG, fMRI, DTI), the type of connectivity (structural or functional), and the process of network reconstruction (analyzing weighted or unweighted networks). Overall, dementia, akin to other brain network diseases such as epilepsy, multiple sclerosis, and schizophrenia, preferentially affects hub areas. Hub changes related to the default mode network or hubs in other association areas are linked to cognitive deficits, particularly those associated with attention, executive function, and working memory. For AD, hubs are also regions with a large amount of amyloid deposition. These changes occur along with reductions in measures of node centrality in AD, particularly in brain regions that can be considered higher-order association areas, such as the temporal lobe, medial parietal, posterior, and anterior cingulate, and medial frontal areas (for a review, see Stam, 2014).

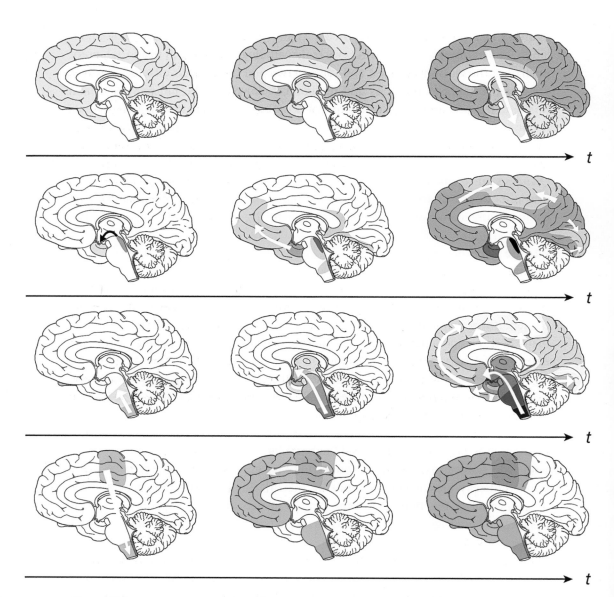

Figure 13.1
Typical spatial progression of protein aggregates in various neurodegenerative diseases (after Jucker and Walker, 2013). Characteristic progression of specific proteinaceous lesions in neurodegenerative diseases over time (*t*, black arrows), inferred from postmortem analyses of brains. Amyloid-β deposits and tau inclusions in brains of patients with Alzheimer's disease (1st and 2nd row), α-synuclein inclusions in brains of patients with Parkinson's disease (3rd row), and TDP-43 inclusions in brains of patients with amyotrophic lateral sclerosis (4th row). Three stages are shown for each disease, with white arrows indicating the putative spread of the lesions.

Other network changes include lower clustering coefficient or reduced local efficiency in AD for EEG, DTI, positron-emission tomography (PET), and magnetoencephalography (MEG) but increased local connectivity for group-level cortical thickness correlations and MRI functional connectivity. Due to a loss of long-distance structural and functional connections, there is also an increase in characteristic path length and decrease of global efficiency. An early example of this is an increase in path length for EEG functional connectivity due to a reduction in long-distance links (Stam et al., 2007). However, reduced path lengths were also reported when, instead of using the absolute path length, a normalized path length, relative to the path length of a random network, is used. In other words, using normalized path length, patient networks are closer to random networks than those of control subjects (Stam, 2014). The modular organization in rs-fMRI functional connectivity in AD patients shows a stronger coupling between frontoparietal and default mode networks, resulting in less segregation of these two networks (Contreras et al., 2019).

For structural connectivity, network changes were linked with behavioral deficits (Lo et al., 2010). For DLB patients, axial diffusivity, radial diffusivity, and mean diffusivity changes were found within the frontothalamic and frontoparietal (precuneus) network whereas, for AD patients, mnemonic pathways (right uncinate fasciculus and right superior temporal gyrus) were affected (Delli Pizzi et al., 2015). These white matter changes in AD were poorly correlated with gray matter changes, indicating that axonal damage and breakdown of oligodendrocytes and myelin could be independent from neuronal loss within the gray matter.

Concerning structural covariance networks, late mild cognitive impairment (MCI) converters to AD, early MCI converters to AD, and AD patients showed a decreased path length and mean clustering coefficient compared with the stable MCI group. These findings suggest that the prodromal and clinical stages of AD are associated with an abnormal network topology (Pereira et al., 2016).

Changes over Time

There is a limited amount of longitudinal data at the moment, but, using large cohorts, it is possible to look at the link between the volume of brain regions at different stages of the disease, including before the onset of dementia. Such studies based on anatomical MRI scans have found an early divergence of AD from the normal aging trajectory before the age of 40 years for the hippocampus and around 40 years for the volume of the lateral ventricles and the amygdala. Further analysis indicates that medial temporal lobe atrophy and ventricular enlargement are two midlife physiopathological events characterizing AD brain (Coupe et al., 2019). Taken together, these findings map onto areas that are already indicated as early starting points for neurodegenerative diseases based on histological changes. Measures based on MRI and fMRI are only some of the many types of biomarkers that change years before symptom onset (see figure 13.2).

Changes in Relation to Disease Molecule Concentration

Changes in connectivity are also linked to a higher concentration of disease molecules, that is, abnormal pathologic proteins. As disease molecules are toxic, the death of projection neurons will also change the strength of structural and functional connections between regions. Strongly connected nodes, based on rs-fMRI functional connectivity, displayed more tau pathology, based on PET imaging, in AD, validating predictions of theories of transneuronal spread but not supporting a role for metabolic demands or deficient trophic support in tau accumulation (Cope et al., 2018).

Changes in Brain Rhythms

These changes in connectivity also lead to changes in brain dynamics. For EEG functional connectivity, delta- and alpha-band activity shows larger changes in AD than in DLB (see, e.g., Babiloni et al., 2018). Looking at microstates, transiently stable topographies of EEG functional connectivity, microstate duration was increased in Lewy body dementia but not in AD or controls. Mean microstate duration was negatively correlated with dynamic functional connectivity between the basal ganglia and thalamic networks and large-scale cortical networks such as visual and motor networks in Lewy body dementia, which might explain how symptoms such as cognitive fluctuation arise (Schumacher et al., 2019).

13.3 Models of Network Changes

Connection Changes due to Altered Processing Demands

Changes in hub regions are linked to cognitive deficits in AD. Furthermore, hubs also show a higher deposition of amyloid. Based on topological and functional considerations, it would be expected that damage at network hubs shows higher clinical behavioral deficits: hubs ensure a relatively low characteristic path length in networks (Kaiser, Martin, et al., 2007), and lesions in hubs, particularly the ones that link different subnetworks, cause the largest disturbances in network organization (see, e.g., Aerts et al., 2016).

Looking at the time course of dementia progression in functional connectivity (rs-fMRI), the posterior default mode network fails *before* measurable amyloid plaques occur and appears to initiate a cascade of connectivity changes that continue as the

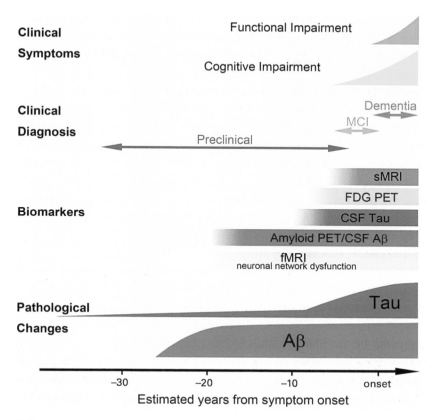

Figure 13.2
Chronological relationships among pathology, clinical symptoms, and biomarkers (adapted from Yoshiyama et al., 2013). Based on biomarker studies, β-amyloid protein accumulation appears to start 20 years before the onset of dementia. MCI, mild cognitive impairment; sMRI, structural magnetic resonance imaging; FDG, 2-[18F]-fluoro-2-deoxy-D-glucose; PET, positron-emission tomography; CSF, cerebrospinal fluid; Aβ, β-amyloid protein; fMRI, functional magnetic resonance imaging.

disease progresses. Moreover, stronger connectivity between the posterior default mode network and (mainly frontal lobe) hubs is associated with more amyloid accumulation (Jones et al., 2016). Jones et al. suggest a cascading network failure that begins in the posterior default mode network and then shifts processing burden to connected network hubs. In other words, observed functional connectivity changes could be due to changing processing demands rather than primarily due to amyloid deposition (Jones et al., 2016). While the majority of models look at the spreading of disease agents (amyloid or tau), functional demands leading to network changes will be important to consider and combine with current models.

Modeling the Spreading of Disease Agents

Disease agents can spread within cells and reach other neurons by diffusing over the axon. Vesicular release can then eject these agents into the synaptic cleft, and in turn these agents can be picked up by the postsynaptic neuron. Agents can also diffuse in the extracellular medium. It is important to keep in mind that diffusion over fiber tracts to remote brain regions can be a slow process. For amyloid molecules, movement from soma to synapse can progress around 1 mm per day (Utton et al., 2005), leading to delays of several weeks for long-distance fiber tracts in humans. Moreover, there will be multiple synapses to cross in order to reach another region. Current models represent a brain region as one network node, which means one step for the propagation of agents. However, in real brain networks, agents that arrive in a region will have to pass multiple synapses before being able to travel to another brain region. Moreover, the number of synapses that need to be crossed and the likelihood of crossing, given by the local connectivity, might differ between brain regions. Altogether, these factors will influence the speed of agent propagation, depending on fiber length and number of synapses along the way, and the amount of agent, given the proportion of agent that is deposited within a region rather than passed on to another connected region. In addition, there may also be local intrinsic facilitators or inhibitors to this spread, influenced by neuronal type, synaptic activity, and other factors. Such considerations are limitations of the models that are described in the following sections.

An early study, based on structural connectivity in healthy subjects, looked at spreading of agents between brain regions (Raj et al., 2012). The model consists of cortical and subcortical nodes, connections between them with connection strength c based on diffusion imaging, and the concentration x of the disease agent at each node. The change of the disease molecule concentration x in a region over time (dx/dt), due to a connected "infected" brain region, is then the product of the concentration difference between both regions $x - x'$ (x' is the concentration of the other connected region), the connection strength between both regions c, and the diffusivity constant β controlling propagation speed. In other words, assuming bidirectional pathways,

$$\frac{dx}{dt} = \beta c(x - x')$$

This describes only concentration changes due to *one* neighbor of a node; the actual change is given by the sum of the concentration changes for all neighbors of a node. Regional atrophy, observable as reductions in MRI volume of regions, was then dependent on the concentration of disease agent for each region. This model found a close resemblance between predicted atrophy and actual T1-weighted MRI volumetrics of Alzheimer's and frontotemporal dementia subjects (Raj et al., 2012). Subsequently, this model was also applied to predict future patterns of regional atrophy and metabolism from baseline regional patterns after the onset of clinical symptoms to the pattern observed 2 to 4 years later (Raj et al., 2015). Here, the model could accurately predict end-of-study regional atrophy and metabolism. This highlights that such models could be useful as a prognostic tool.

A study in mice with the olfactory bulb as a starting point of α-synuclein found that a computational model that assumes spreading through fiber tracts better reproduced the experimentally observed spreading pattern than a model based on proximity-based propagation by diffusion (Mezias et al., 2019). Interestingly, this study indicates that the preferred direction of α-synuclein inclusion propagation is initially retrograde and switches to anterograde propagation several months after injection.

In contrast, a study on spreading in humans found that typical degeneration patterns in various prion-like diseases could also be reproduced in a model that combines spreading through fiber tracts with spreading through the extracellular medium within the gray and white matter (Weickenmeier et al., 2018). Using a mechanical finite-element model, cell atrophy leads to morphological changes that replicate changes observed in patients: pronounced hippocampal atrophy, ventricular enlargement, and a widening of the cortical sulci. When comparing the roles of extracellular and fiber tract (transneuronal) spreading, Kim et al. (2019) reported a larger role of transneuronal propagation. Note, however, that this study used resting-state functional connectivity as a proxy of fiber tract strength. Real structural connections between brain regions might, however, be imbalanced between both directions or even unidirectional. Altogether, such models will be useful when one aims to reproduce the morphological changes that are observed at different stages of dementia and taking into account different types of dementia with potentially different starting regions.

Such progression models reporting the sequence and extent of disease agent propagation can help to predict future disease trajectories. In a longitudinal study with 1,000 subjects, stable versus declining clinical symptom trajectories could be distinguished, and, given baseline and 1-year follow-up, the subsequent disease state could be predicted (Bhagwat et al., 2018).

Activity-Dependent Changes
We mentioned that neurotransmitter release at the chemical synapse, including the unwanted payload of disease agents within released vesicles, is needed for transneuronal propagation. Experimental studies show that excessive neuronal activity leads to an increase in amyloid deposition. However, activity was not part of the previously discussed models.

With cortical brain regions represented as neural masses, each describing the average activity (spike density and spectral power) of excitatory and inhibitory neurons, and connections between regions based on DTI, structural and functional network hubs showed higher average activity (de Haan et al., 2012). "Activity-dependent degeneration" was simulated by lowering synaptic strength as a function of the spike density of the main excitatory neurons. Observed transient increases in spike density and functional connectivity are comparable to those seen in MCI patients. This suggests that neural activity can be involved in degeneration and could explain why network hubs, which are overall more active, are predominantly affected (de Haan et al., 2012).

In a subsequent computational study, different interventions were tested to slow down the activity-dependent degeneration, targeting excitatory, inhibitory, or both populations of neurons (de Haan et al., 2017). Surprisingly, the most successful strategy was a selective stimulation of all *excitatory* neurons in the network. This highlights potential benefits of brain stimulation, or of other interventions that increase excitatory neuron activity, in slowing down disease progression (see also chapter 15).

Modeling Changes in Brain Connectivity
Atrophy in the previous models resulted in loss of gray matter, leading to volume and morphological changes. However, the death of projection neurons will also lead to fewer axons passing through a fiber tract and therefore a lower connection strength as observed through structural connectivity.

In the following model (Peraza et al., 2019), the strength of a connection w will decrease depending on its current strength w, the concentration x of the disease agent in both connected regions i and j, as well as on the rate of connection strength change α:

$$\frac{dw_{ij}}{dt} = -\alpha w_{ij}(x_i + x_j)$$

In addition to the disease-agent-related change in connection strength, there is also a generic decrease of connection strength as part of the aging process. Note that this and the other models presented here do not include clearance, that is, the removal of a certain amount of the disease agent over time (for more information on including clearance within models, see Georgiadis et al., 2018).

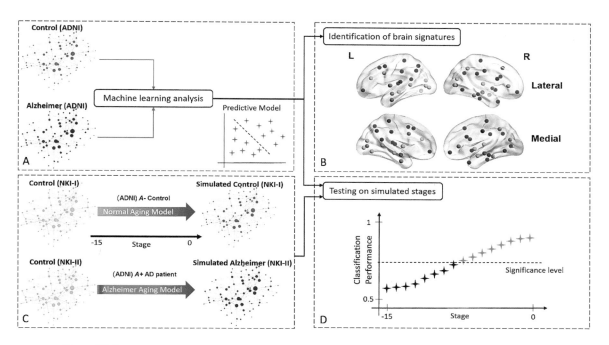

Figure 13.3
Simulating healthy aging and progression toward dementia. (A) In the first part of the work, Peraza and colleagues generated a predictive model based on measures of complex networks to classify between a cohort of patients with dementia and a cohort of matched healthy controls. (B) They then extracted informed brain signatures enabling diagnosis. (C) In the second part, they applied dynamical models to emulate changes in structural connectivity caused by normal aging, on the one hand, and degradation of structural connections caused by Alzheimer's disease on the other hand. (D) They finally explored when the relevant features and brain signatures associated with dementia begin to be evident in the simulated progression process. Significance level represents the minimum value from which classification performance is statistically significant. ADNI = Alzheimer's Disease Neuroimaging Initiative, NKI = Nathan Kline Institute, AD = Alzheimer's Disease (after Peraza et al., 2019).

For modeling dementia progression, each brain region can be a seed node for the initial deposition of disease agents. However, starting with structural connectivity in healthy younger subjects, only a few seed regions led to simulated progressions that matched the structural connectivity observed in older dementia patients. The most plausible scenario was when the disease is initiated in the hippocampus, as well as in the entorhinal cortex and the amygdala (Peraza et al., 2019). The same study reported also that this scenario matches experimental data on atrophy and functional disruption that identified the hippocampus and the entorhinal cortex as being involved at very early stages of AD.

When simulations with and without disease agents are compared, characteristic topological changes occur before the final stage that represents the onset of clinical symptoms (see figure 13.3). These early biomarkers of progression toward dementia are mainly network centrality measures largely associated with medial temporal and subcortical brain regions, as well as posterior structures of the default mode network and occipital areas (Peraza et al., 2019). This highlights the possibility that neural structure may be significantly compromised before cognitive deficits become evident enough for a clinical diagnosis.

For Parkinson's disease, a network diffusion model was compared to the data of 232 patients (Pandya et al., 2019). When different potential spreading starting points were looked at, the substantia nigra was the most likely seed region. Spreading based on structural connectivity provided a better predictor than spreading based on distance or performed randomly and also a better predictor than regional gene expression.

Altogether, computational models of dementia progression based on the connectome are an approach to discover risk factors and biomarkers of brain network diseases and potentially stratify disease cohorts. They can potentially be used to predict the progression toward dementia and the progression after the onset of clinical symptoms. Future clinical studies, including ones with larger longitudinal cohorts and multiple data sources ranging from MRI/EEG to blood samples and cerebrospinal fluid, will be needed to assess the clinical application of these models.

14 Recovery from Injury

14.1 Overview

Connectomes can also change because of tissue damage. Damage can occur inside the brain, for example, after stroke, or affect the sensory input that arrives within the brain, for example, after injury in the periphery. Lesions of brain tissue can be focal, affecting only one region or fiber tract, or diffuse, affecting multiple network nodes or edges. Some changes are deliberate, such as in the surgical resection, or in situ destruction, of brain tissue in cases of brain tumors or medically intractable epilepsy (for a broader discussion of brain damage and recovery, see Finger and Stein, 1982).

To what extent the brain can recover from lesions depends not only on the lesion itself but also on the capability of one's brain network to functionally compensate for the affected brain tissue. The concept of degeneracy describes the ability of other brain regions to perform the same function or to yield the same output as the lesioned regions (Tononi et al., 1999). A measure for degeneracy is its order (Price and Friston, 2002); that means the minimum number of regions that need to be removed before a function is lost (first-order: one region, second-order: two regions, etc.). If a function can be retained after a lesion, even though function performance may be reduced, this is known as functional compensation.

The ability to functionally compensate for lesions can be understood with a view toward network connections or network nodes. As we will see later in this chapter, nodes that already have incoming and outgoing connections that are similar to those of the lesioned regions have a better capacity to (partially) take over the function. This is less likely if all remaining regions have a very different connectivity profile than the lesioned regions. From the point of view of network connections, we might look at alternative pathways: if a node (or edge) on a path between two regions is damaged, how many remaining paths exist between the two regions?

Finally, networks are rewiring as a consequence of the lesion. While it was initially believed that major rewiring would only be possible when lesions occur early during development, such changes have now also been observed at the adult stage in mice (Keck et al., 2008). These changes can be beneficial, to compensate for the function of lost tissue, but can also lead to further deterioration of brain function. After traumatic brain injury or the growth of brain tumors, surrounding tissue can undergo changes that ultimately lead to the rise of epileptic seizures later on, sometimes years after the initial injury. In this chapter we will first observe how brain networks change after lesions and then look at a conceptual framework for modeling how and to what extent brain function can be recovered.

14.2 Examples of Reorganization after Lesions

Modeling Lesion Effects

Biological systems, unlike most human-made networks, can be remarkably robust toward the elimination of components. For yeast, single-gene knockouts are nonlethal in 70% of the cases, leading to reduced metabolic activity but not affecting survival or reproduction of the organism (Wagner, 2000). It is thought that such network robustness is due both to redundancy (same genes at different locations of the genome) and alternative metabolic pathways (e.g., anaerobe and aerobe for some bacteria).

Within a theoretical framework, there are two ways to model the immediate effect of network damage: running a simulation of network function or dynamics, or observing the change in topological network features. An often-used topological feature is the ability of information to pass through the network as measured by the characteristic path length. When one is comparing the average number of connections to cross to go from node to another of the lesioned network with the value for the unlesioned original network, a larger increase in average path length is thought to indicate greater damage for the system and potentially a larger functional impact. This idea relies on several assumptions. First, it assumes that information can pass over any connection. Therefore, there is no such thing as a fiber tract that is limited to visual or auditory information. Second, it is assumed that only the number of steps—one for each passed connection—is important. In other words, the spatial length of a connection, and therefore the delay in propagating information, is not taken into account. Despite these limitations, such an analysis gives us a first estimate of what lesion targets will lead to a major decrease in network function.

Looking at the removal of nodes or edges in the brain networks of cat and macaque, the removal of hub nodes led to the largest increase in characteristic path length.

Compared to random removal of nodes, the removal of a few hub nodes already leads to a 50% increase in the length of shortest paths (Kaiser, Martin, et al., 2007). Hub nodes are crucial shortcuts in connecting different parts of the network. In the same way, removing connections that are involved in many shortest paths—the ones with high edge betweenness—also leads to a severe increase in shortest path lengths. Altogether, the reaction to damage of brain networks is similar to benchmark networks with hubs but different from small-world or random benchmark networks. This indicates that the hub architecture of connectomes is crucial for the outcome of network lesions.

Alternatively, the effect of lesions on the dynamics of network activity can be simulated. Looking at simulated "resting-state" neural activity based on an underlying human brain structural network, node removal led to activity changes not only in nodes to which lesioned nodes were previously connected but also in distant nodes, often in the contralateral hemisphere (Alstott et al., 2009). Stronger and more widely distributed effects were observed for lesions along the cortical midline and near the temporoparietal junction. Damage to these regions is also known to lead to severe cognitive deficits in humans. Interestingly, the dynamic effects of lesions were not always predicted from the topological effects that we discussed previously. Degree and strength were weak predictors while number of connections between lesion site and the rest of the brain as well as how much the lesion increased the path length of the remaining network performed better. The most robust prediction was made by the extent to which the lesion damaged the default mode network.

Rewiring after Lesions That Occur Early On
When looking at functional recovery after lesions, there are three different stages: the acute stage within the first week of the lesion, the subacute period where some functionality returns, and the chronic period where performance stabilizes. An important consideration for the final outcome of lesions during the chronic period is the extent to which the network can rewire to try to recover lost brain function. In general, brain plasticity is thought to be higher during early stages of brain development. The loss of functional brain tissue through lesion or surgery can be more easily compensated for if it happens during earlier stages of brain development. For example, there is a greater functional recovery after temporal lobe epilepsy surgery in children than in adults (Gleissner et al., 2005).

An example of recovery from early lesions is the outcome of the removal of the entire primary visual cortex (areas 17, 18, and 19) in newborn kittens. While such damage cannot be compensated for in adults, for kittens another region, the cat's posteromedial lateral suprasylvian (PMLS) area of the cortex, is able to take over the function of the removed visual cortex (Spear et al., 1988). This is related to an underlying change in the synaptic organization after the lesion; for example, the percentage of direction-sensitive cells in PMLS decreases from nearly 80% in normal cats to about 20% after the lesion. While some visual functions are already, to a lower extent, present in PMLS before the lesion, such as ocular dominance, others, such as direction sensitivity and orientation selectivity, arise de novo after the lesion. Recovery does not rely on the remaining contralateral visual cortex as compensation also occurred for bilateral removal of areas 17, 18, and 19 (Guido et al., 1990).

Rewiring at Adult Stages

Recovery and rewiring are, to some extent, also possible after brain maturation. For motor recovery after stroke, for example, motor function shows early recovery but then plateaus after 3 to 6 months and around half of the patients retain a paralysis of one side of the body afterward (Bundy and Nudo, 2019).

Because of the ability to have defined focal lesions, there are many studies in animals about changes after lesions. For nonhuman primates, there is increased functional connectivity 3 to 4 months after a lesion to the primary motor cortex, M1 (Dancause et al., 2005). Moreover, M1 injury results in axonal sprouting near the ischemic injury and the establishment of novel connections within a distant target, supporting the idea that cortical areas distant from the injury undergo major neuroanatomical reorganization. Studies in rats have shown that effects are also visible in the contralesional hemisphere following rehabilitative training, indicating that both hemispheres interact during the process. In the four lesioned monkeys, Dancause et al. found novel connections from the ventral premotor cortex to a part of the somatosensory cortex; these connections were absent in all four control animals. Inspection of the orientation of labeled fibers revealed sharp changes in direction as fibers approached the lesion site, diverting to innervate the somatosensory cortex instead. The somatosensory cortex projects directly to the spinal cord, and so this rerouting may provide an alternative route for premotor cortex to innervate the spinal cord following damage to the primary motor cortex. These results raise the exciting possibility that damage can provoke the diversion of fibers and growth of novel brain connections that could provide a substrate for functional recovery (Johansen-Berg, 2007).

For rodents, axonal sprouting forms new intracortical connections within perilesional areas of the ipsilesional hemisphere as well as new corticospinal and corticorubral connections stemming from both the ipsilesional and contralesional hemispheres. Despite the structural and functional changes that are observed after lesions, it is important to remember that the behavioral performance improves earlier with receptive field changes following with a certain delay. Finally, note that in rats and nonhuman primates, motor recovery might be easier to achieve than in humans because of a larger extent of alternative pathways given by the rubrospinal and reticulospinal tracts (Bundy and Nudo, 2019).

The effects of lesions have also been observed with respect to structural connectivity changes in human patients (Johansen-Berg et al., 2010). White matter changes, detected by diffusion MRI, in pathways related to motor execution and motor learning can be used to predict recovery following stroke or response to intervention. For MEG functional connectivity changes after brain injury, a loss of delta- and theta-based connectivity and, conversely, an increase in alpha- and beta-band-based connectivity were found (Castellanos et al., 2010). The same study found that network reorganization and cognitive recovery were linked in that the reduction of delta-band-based connections and the increment of those based on alpha band correlated with Verbal Fluency test (phonetic verbal fluency total score), as well as Perceptual Organization and Working Memory Indexes, respectively. Also, changes in connectivity values based on theta and beta bands correlated with the Patient Competency Rating Scale.

14.3 Mechanisms of Functional Recovery

Homeostasis: How Does Activity-Dependent Structural Plasticity Change Neuronal Circuits?

The concept of homeostatic structural plasticity (Butz et al., 2008; Butz et al., 2014) can be applied to understand what happens after lesions leading to loss of sensory (cortical deafferentation) or loss of corticocortical input. While functional plasticity describes the changes in synaptic weights between connected neurons, structural plasticity describes the formation or removal of a connection between neurons (Johansen-Berg, 2007). Structural plasticity goes beyond changes at the synapse itself and can encompass forming or breaking of motile spines and rerouting of axonal branches in the developing and adult brain (see figure 14.1). Also, structural plasticity can occur during brain maturation as well as after lesions (Butz, Wörgötter, and van Ooyen, 2009).

If tissue is damaged, many of the remaining neurons, especially adjacent to the lesion site, will have a lower number of incoming connections. In order to still meet a set point of firing activity, new spines and boutons will be created.

An example of structural plasticity at the adult stage is the reorganization of the visual cortex after the loss of input as a result of focal retinal lesions (Butz and van Ooyen, 2013). After the lesion, the turnover of dendritic spines increases, in particular

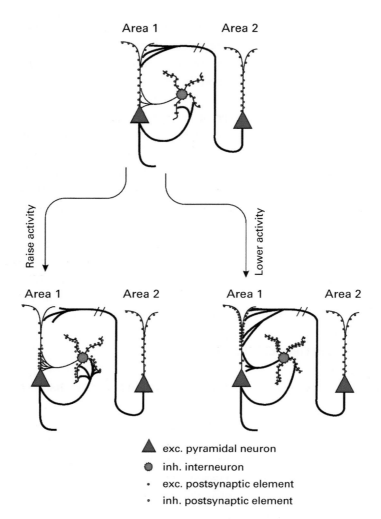

Figure 14.1
Activity-dependent structural plasticity of neuronal networks. If network activity is raised, postsynaptic neurons reduce excitatory postsynaptic elements to lower excitation and offer more inhibitory postsynaptic elements to increase the chance to receive more inhibition. At the same time, they respond with axonal sprouting (axons are drawn in bold). Sprouting axons of inhibitory interneurons serve the demand of highly activated neurons so that an increasing feedback inhibition brings the network in balance again. The opposite happens if activities are too low. Then, inhibition is reduced, and the amount of excitatory postsynaptic elements increases. As activity is needed for axonal sprouting, compensation for low network activity requires the supply of additional axonal offers from different areas that are highly activated (adapted from Butz, Wörgötter, and van Ooyen, 2009).

in the center of the lesion projection zone, while the number of axonal boutons initially increases followed by a strong reduction. Furthermore, new horizontal connections cause a retinotopic remapping. Using a computational model of this process, the neuron and network changes within the visual cortex after focal retinal lesions can be reproduced (Butz and van Ooyen, 2013). In the model, as in the cortex, the area that showed the greatest increase in the turnover of dendritic spines was the center of the lesion projection zone, while axonal boutons displayed a marked overshoot followed by pruning. In addition, the decrease in external input was compensated for by the formation of new horizontal connections, which caused retinotopic remapping.

Using a computer model with an activity-dependent rewiring rule, there are also distinct topological changes after lesions (Rubinov et al., 2009). Activity-dependent plasticity leads to a network that is robust to random removal of nodes, showing preserved small-world features. On the other hand, targeted removal, aiming for nodes with high node centrality, leads to a drop in reachability and global efficiency, resulting in a loss of the small-world architecture. However, increasing neurogenesis, adding new nodes with random connectivity, can compensate for this targeted removal even when the growth process is slower than the node removal process.

Functional Compensation: Which Regions Can Take Over the Function?

If one region loses its function, another area with similar function or with the same function operating to a lesser extent could overtake the function of the affected region. Compensation without gross rewiring of brain pathways is referred to as recovery whereas compensation at earlier developmental stages caused by unmasking of existing pathways or altered development of brain pathways is called sparing of functions (Payne and Lomber, 2001).

As discussed earlier, a loss of the visual cortex (areas 17, 18, and 19) in the first weeks after birth in kittens can be compensated for by the PMLS. After lesion of the visual cortex an encoding of direction sensitivity and ocular dominance was found by electrophysiology in PMLS. While these functions are preexisting to a lesser extent before the lesion, others such as orientation selectivity develop de novo (Spear et al., 1988). A necessity for the compensating area seems to be similar incoming and outgoing connections. For example, a region that compensates for the loss of parts of the visual cortex should get incoming visual information (e.g., from the LGN or the superior colliculus).

To measure the similarity based on the connectivity of the whole network, nonmetric multidimensional scaling (NMDS) is used. An advantage of NMDS over the matching index or Jaccard index, which measure how similar neighbors of a node are, is

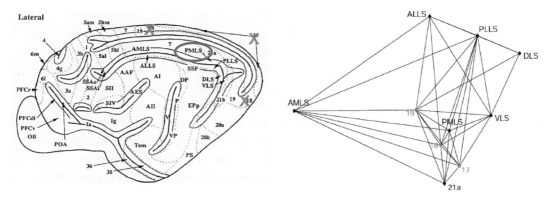

Figure 14.2
Functional compensation facilitated by connectivity overlap. Left: Lateral view of the cat cortex (from Scannell et al., 1995). The location of the lesioned visual cortex (areas 17, 18, and 19) is shown. Posteromedial lateral suprasylvian (PMLS) is the region that exhibits functional compensation for the damaged visual areas. Right: Two-dimensional projection of the location of the regions in the nonmetric multidimensional scaling output space. The connections between visual regions are shown. PMLS is near to the removed visual areas 17, 18, and 19. The closest other candidate regions are the ventrolateral suprasylvian area (VLS) and area 21a.

that NMDS also looks at indirect connections, including neighbors of neighbors and beyond. Multidimensional scaling means that the network space, which initially has as many dimensions as there are nodes, is reduced to two- or three-dimensional space. Nodes that are nearby one another in this reduced space have similar connectivity patterns.

Looking at structural connectivity in the cat cerebral cortex, amygdala, and hippocampus (Scannell et al., 1995, plus additional data), NMDS yields PMLS on the first place being closest to the removed regions indicating the most similar connectivity profile. The average NMDS distance between PMLS and the removed areas 17, 18, and 19 has been 0.33 (see figure 14.2) whereas the distance for the 61 other candidate regions was at least twice as large. In addition to identifying the region that is involved in functional compensation, the distance to the removed regions in NMDS space might be an indicator of the extent to which function can be recovered.

Cognitive Reserve: How Likely Is Recovery?

Cognitive reserve refers to the brain's ability to cope with increasing brain damage or age-related degeneration while still functioning appropriately. This concept originates from the repeated observation of the inconsistency between the severity of brain pathology and the clinical deterioration. For example, postmortem studies (Katzman et al., 1988) have shown that subjects might have high depositions of misfolded proteins without showing symptoms of Alzheimer's disease (AD).

An individual's cognitive reserve can be influenced by occupational attainment, leisure activities at old age, and education (Katzman et al., 1993). For example, individuals with lower education levels had higher risk of developing AD, while individuals with higher education levels showed less chance of developing AD; however, they showed a more rapid decline of cognitive function when they got AD. This difference is thought to be due to the ability to maximize cognitive performance through differential recruitment of brain resources or due to the use of alternate networks during the course of brain degeneration.

A network measure that has been applied as a marker for cognitive reserve is network flow (Yoo et al., 2015). For a binary graph, the maximum flow value is equal to the number of edge-disjoint paths between source and destination nodes in the graph. Represented by the maximum flow values, the number of edge-disjoint paths between a pair of nodes in each subject could be correlated with education levels in a positive or negative way. A subnetwork for the normal control group, centered at the left supramarginal gyrus, has a positive correlation between the maximum flow values and education levels. Another subnetwork for an AD patient group, centered at the left middle frontal gyrus, has a negative correlation between the maximum flow values and education levels. Both maximum flow subnetworks include a hub region, the precuneus. For the subnetworks of the AD group, the robustness of connectivity had a negative correlation with education levels. This finding supports the hypothesis that cognitive functions are more severely impaired in AD patients with higher education levels and that network flow can be a measure to assess cognitive reserve (Yoo et al., 2015).

Timing of Lesions: Some Times Are More Critical than Others

As we saw before, there often seems to be a higher capacity for rewiring and functional compensation during brain maturation compared to when lesions occur at the adult stage. The timing of lesions can be observed in a computational model of the primary sensorimotor cortex. This model includes self-organizing maps in the motor system that are used to control arm movement. For the adult stage, cortical map reorganization after a simulated infarct depended on perilesion excitability: lesions to cortical regions with increased perilesion excitability were associated with a remapping of the lesioned area into the immediate perilesion cortex. On the other hand, when lesions caused a perilesion zone of decreased activity to appear, this zone enlarged and intensified with time, with loss of the perilesion map (Goodall et al., 1997).

For lesions during brain maturation, lesions were applied during the training stages of the neural network; that means during the time when cortical maps were forming (Varier et al., 2011). After the lesion, the ongoing map formation and network function, in terms of controlling arm movement, were monitored as "development" progressed to completion. Activities in recovering systems injured at an early stage show changes that emerge after an asymptomatic interval. Early injuries cause qualitative changes in system behavior that emerge after a delay during which the effects of the injury are latent.

In addition to the timing of the lesion, it also made a difference whether lesions were focal, affecting an adjacent patch of the map, or diffuse, removing tissue at many distributed spots on the map (Varier et al., 2011). Lesion effects were greater for larger, earlier, and distributed (multifocal or diffuse) lesions. On the other hand, the mature system is relatively robust, particularly to focal injury. A final point on timing, as shown in experiments in cats, is that gradual lesions where tissue is removed in several steps with surgeries 1 to 2 weeks apart are more easily compensated for than removing the same total amount of tissue in one stage (Finger and Stein, 1982).

In summary, the connectome shows changes after lesions, both adjacent to the site of injury and also in faraway regions. Some of these changes are positive, leading to functional compensation, while others have a negative impact on recovery. In order to recover, neural systems can rely on existing circuits using functional compensation or establish new connections using activity-dependent synaptic rewiring and axonal sprouting (for more information, see van Ooyen and Butz-Ostendorf, 2017). Computational models can be used to assess the role of lesion size, timing, and target, but further longitudinal studies will be needed to better understand the mechanisms of recovery.

15 Brain Stimulation Effects

15.1 Overview

Changing brain activity, either to tackle brain disorders or to enhance cognitive functions in healthy subjects, is an important goal in neuroscience research. While drugs can change brain activity, once they pass the blood-brain barrier they diffuse through the whole brain, leading to both effects and side effects. Brain stimulation, on the other hand, has the potential to be more focal about which parts of the brain are targeted. Brain stimulation or neuromodulation technologies are divided into noninvasive, invasive (devices are placed inside the brain), and semi-invasive (devices are inside the body but outside the brain).

There are also different kinds of stimulation. Open-loop stimulation is continuously stimulating neural networks, regardless of the state of the brain. Behaviorally triggered stimulation only occurs related to certain events. For example, stimulation before a task can improve task performance. For closed-loop stimulation, the stimulation depends on the current state of the system: brain activity is measured at one site, whether and how to stimulate is determined, and stimulation is then applied to either the measurement site or a different site. Such a stimulation requires continuous monitoring and an adjustment of parameters to achieve optimal results for individual patients.

There are also different modalities for brain stimulation (see figure 15.1). Noninvasive electric stimulation has been used in varying forms for centuries, with reports from the ancient world of electric fish used to provide stimulation for headaches. Today, the use of noninvasive electrical stimulation is mainly focused on transcranial current stimulation (TCS) and transcranial magnetic stimulation (TMS), and it is used for treatment as well as research. TCS involves placing electrodes over the scalp and channeling a current flow from a positive "anode" electrode to the negative "cathode." TCS is essentially modulating the excitability of the stimulated tissue (Nitsche et al.,

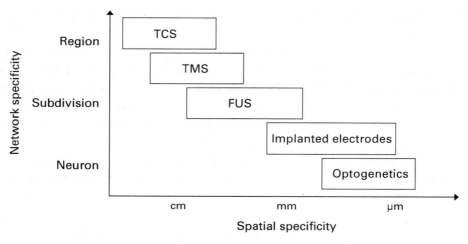

Figure 15.1
Spatial and network specificity of different kinds of brain stimulation (schematic overview). For transcranial current stimulation (TCS), current could also have effects outside the target tissue as it passes between electrodes. For transcranial magnetic stimulation (TMS), effects are more localized and could activate a small patch of tissue on the cortical surface. For focused ultrasound stimulation (FUS), effects could be targeted to a small population of neurons with potentially direct effects within a few mm^3. For implanted electrodes, effects can change the local field potential and, for nearby neurons, influence their firing activity. For optogenetics, specific types of neurons can be targeted with a direct effect on firing as a result of a light stimulus. Note that implanted electrodes as well as optogenetics, with the injection of viruses and the implantation of an LED, are invasive techniques.

2008). The type of current can be varied, with the most studied being transcranial direct current stimulation (tDCS), where the current is ramped up to a steady level, which is then applied for the duration of the stimulation protocol. Other variants, which are growing in popularity, include transcranial random noise stimulation and transcranial alternating current stimulation (tACS). A more recent type of noninvasive electrical stimulation is temporal interference, where multiple electric fields with different frequencies are applied and where, at the location of maximum interference, neural oscillations become entrained to the difference in the frequencies (Grossman et al., 2017). For example, application of fields with 1,000 Hz and 1,010 Hz would lead to entrainment in the 10 Hz frequency band for the location of maximum interference between the fields. In a study with rodents, temporal interference was not only able to reach deep-brain structures such as the hippocampus without physiological effects in intermediate tissue but could also modulate brain activity within a distinct frequency band (Grossman et al., 2017).

TMS involves inducing currents in the brain by applying magnetic field changes. Unlike with electrical stimulation, with TMS the skull does not act as a barrier to the stimulation as magnetic fields are not restricted by it. TMS pulses produce a large enough electric current to elicit action potentials in the underlying tissue and have been used to create "virtual lesions" where ongoing brain activity is disrupted for short durations to discern the function of different regions.

Focused ultrasound neuromodulation operates in the frequency range above human hearing. Tissue is affected by acoustic pressure waves that, for high intensity, can lead to increases in temperature and that are used for tissue ablation—for example, for patients with brain cancer or Parkinson's disease. For lower intensities and shorter exposure times, mechanical waves can modulate cell membrane features that affect the likelihood of neurons firing (Jun, 2012). Among the underlying factors that are being discussed are these: cavitation, leading to small gaps between the two layers of the cell membrane and resulting changes in membrane capacitance (Lemaire et al., 2019), mechanical effects leading to the opening of ion channels (Kubanek et al., 2018), and thermal effects. The ability to, depending on the stimulation protocol, either increase or decrease neural activity and to reach deep-brain structures has several potential clinical applications (Leinenga et al., 2016). Stimulation effects can modulate activity in the human thalamus (Legon et al., 2018) and, as shown for macaques, can last for more than an hour (Verhagen et al., 2019).

For invasive techniques, implanted electrodes are also able to directly target deep-brain structures. Such deep-brain stimulation is used in Parkinson's disease to reduce tremor but also in obsessive-compulsive disorder, dystonia, and epilepsy. Applications in other brain disorders are currently being tested. As an invasive technique, there are risks related to the performance of surgery, and therefore this approach is unsuitable for healthy subjects and only a last resort for patients if standard medical treatments are ineffective.

Another invasive technique is optogenetic stimulation that uses light sensitive ion channels to manipulate the membrane potential and activity of a target cell. The advantage is that selective cell types can be targeted (using cell-specific promoters) to express these ion channels. This leads to a great level of fine-grained control of the stimulus, which is not achievable by any conventional brain stimulation devices. Because of these advantages, optogenetics has become a vital tool for studying and controlling neural circuits. Optogenetics as a therapeutic tool is still in its infancy, but promising studies are paving the way for optogenetics to become a clinically accepted tool.

15.2 Effects of Brain Stimulation

Local Changes after Brain Stimulation

How does brain stimulation affect neural networks? For electrical direct current stimulation, in vitro experiments in rat hippocampus have shown persistent changes that lasted more than 10 minutes after stimulation (Reato et al., 2015). The underlying changes in excitation and inhibition could be due to several mechanisms: (1) activity-dependent brain-derived neurotrophic factor could increasingly be released, in turn increasing acetylcholine concentration, and due to the longer release time effects could last beyond the stimulation period; (2) spike-time-dependent plasticity could take place, leading to longer-term effects; and (3) stimulation-induced slow changes in neuromodulator release could lead to slow changes in neuronal membrane excitability. Brain stimulation engages the glutamatergic system in humans and animals, was shown to interact with the GABAergic systems in animals, and can lead to long-term plasticity (Nitsche et al., 2012).

For invasive stimulation, a study in humans using intracranial electrical stimulation (single-pulse corticocortical evoked potentials) to monitor excitability showed that some regions showed both potentiation and suppression of excitability that persisted for at least 10 minutes (Keller et al., 2018). Such regions with effects beyond the time of stimulation were spatially close to the stimulated site and had high-amplitude evoked potentials during the stimulation. Another study (Mohan et al., 2019) has shown that effects of intracranial electrical brain stimulation in humans are frequency dependent, where stimulation at low frequencies suppresses and stimulation at high frequencies is more likely to activate brain tissue.

Changes in Brain Connectivity

Stimulation can also change the strength of functional connections between brain regions. In a study with implanted electrocorticography (ECoG) electrodes in human epilepsy patients (Khambhati et al., 2019), the coherence pattern between electrodes before and after electrical stimulation was measured. Stimulation had two potential outcomes, either distributed changes in functional connectivity across the whole ECoG grid or localized changes involving fewer electrodes. Which of the two outcomes occurred was linked to structural connectivity: for stimulated regions with weak structural connectivity there was preferentially a modulation of the functional hubness of downstream brain regions and a large change in the dynamical brain state. This, along with studies in healthy subjects (Li et al., 2019), indicates that information about structural connectivity can be used to choose stimulation targets in order to get the desired outcome on local or large-scale functional connectivity changes.

The effects on functional connectivity can be modulated by pairwise stimulation and changing the relation between the stimulation frequency and the frequency of ongoing brain oscillations. tDCS anodal stimulation of the primary motor cortex (M1), combined with cathodal stimulation of the contralateral frontopolar cortex, results in increased EEG functional connectivity within premotor, motor, and sensorimotor areas of the stimulated hemisphere during motor activity in the 60 to 90 Hz frequency range (Polania et al., 2011). Concerning the relation to ongoing oscillations, in this case in the gamma band, in-phase high-definition tACS stimulation of the parieto-occipital cortex enhanced synchronization whereas antiphase stimulation impaired functional coupling (Helfrich et al., 2014). Furthermore, the increases in the gamma band during an ambiguous motion task were linked to decreases in the alpha band.

Connectivity changes can be an important component for assessing the effects of brain stimulation in patients. In a study of drug-resistant focal epilepsy patients, cathodal tDCS led to increased EEG functional connectivity of the epileptic focus which was correlated with a reduced frequency of epileptic seizures (Tecchio et al., 2018). Reduced seizure frequency following tDCS intervention could also be observed in focal *status epilepticus* in POLG-related mitochondrial disease patients (Ng et al., 2018).

Improvement of Motor Recovery after Lesions

In the previous chapter, we saw that lesions result in structural rewiring of the remaining tissue with the potential to partially recover initially lost brain functions. Brain stimulation has been applied to facilitate such recovery (Bundy and Nudo, 2019). For example, open-loop 100 Hz epidural cortical stimulation to the perilesional areas of rats, when paired with rehabilitation, improves motor recovery relative to rehabilitation alone. Furthermore, vagus nerve stimulation has been considered to increase neural plasticity. Finally, in a rodent model of motor injury, lesioning the caudal forelimb area, triggering electrical stimulation in the primary sensory cortex based on the timing of action potentials that were recorded from the remaining rostral forelimb area, improves the performance of a skilled pellet retrieval task when compared either with an open-loop sham stimulation protocol or with a control group receiving no stimulation (Bundy and Nudo, 2019).

In a computational study of synaptic plasticity after lesions, paused stimulation of the networks was found to be much more effective in promoting reorganization than continuous stimulation (Butz, van Ooyen, and Wörgötter, 2009). This was supposed to be the case due to neurons quickly adapting to stimulation, whereas pauses prevent a saturation of the positive stimulation effect. Indeed, the spacing of stimulation, with gaps of more than 10 minutes between subsequent stimulation within a session, has been shown to be linked with prolonged plasticity and behavioral aftereffects in experimental animal and human studies (Goldsworthy et al., 2015). Future studies will be needed to show how behaviorally triggered or closed-loop pulsed stimulation could facilitate rehabilitation in human patients.

Improving Cognition

Stimulation techniques have been shown to increase performance in cognitive tasks (Nitsche et al., 2003; Nitsche et al., 2008) and have shown potential to induce behavioral changes in clinical populations (Kekic et al., 2016). tDCS and TMS have already shown promising results, particularly in alleviating symptoms of hemineglect and aphasia. Dementia patients show deficits in working memory with changes in theta–gamma phase–amplitude coupling in the temporal cortex and theta phase synchronization across the frontotemporal cortex. Using high-definition tACS with a frequency of 8 Hz in older adults (60–76 years), an increase in synchronization and the sender–receiver relationship of information flow within and between frontotemporal regions could be observed (Reinhart and Nguyen, 2019). As a result, working memory performance improved, an effect that lasted for more than 50 minutes beyond the time of stimulation.

Brain stimulation can also improve the mood of patients, and repetitive TMS is used for some cases of major depression. Another technique for treating major depression and obsessive-compulsive disorder is deep-brain stimulation with implanted electrodes. It was found that ventral internal capsule/ventral striatum deep-brain stimulation enhances cognitive control (Widge et al., 2019). Furthermore, enhanced control is correlated with the expected prefrontal cortex theta oscillations, and the increased theta power is in turn correlated with clinical recovery. These distinct changes in network dynamics might therefore be a more objective measure of a stimulation effect than subjective mood assessments by patients.

The effect of stimulation on cognition also depends on the structural connectivity of an individual. One study observed electrical stimulation of the saliency network, a part of the connectome that is crucial for cognitive control such as response inhibition (Li et al., 2019). Anodal tDCS stimulation of the saliency network, targeting the right inferior frontal gyrus/anterior insula cortex, improved response inhibition. However, subjects with high fractional anisotropy within the saliency network showed improved cognitive control while those with low fractional anisotropy did not. This highlights that connectome features could be used to predict the effectiveness of brain stimulation interventions.

A cautious note on improvements after brain stimulation is the need to look at all effects of an intervention. Paradoxical improvements, where the desired function improves but a different function deteriorates, can arise when stimulation disrupts a process that has a suppressive effect on the process of interest (Bestmann et al., 2015). A broad array of behavioral assays is therefore needed to notice both desired effects as well as negative side effects.

15.3 Models of Stimulation-Induced Connectome Changes

Modeling Stimulation Strength in Target Regions

The first step in assessing the effects of brain stimulation is to determine what the brain location of maximal stimulation is and how much stimulation energy is delivered. For TMS this is straightforward as a magnetic pulse will be delivered to the tissue underneath the stimulation coils and strength of the pulse is not affected by the tissue that is between the stimulation device and the target region. Instead, the TMS-induced electrical field is dependent on the coil position, shape, orientation, and pulse strength.

For electrical and focused ultrasound neuromodulation, however, the extent of and the type of material that is between the stimulation device and the target region will determine how much of the initial energy will arrive at the target. For example, the skull absorbs most of the energy delivered by focused ultrasound, and tissue properties influence the flow of electricity for electrical stimulation. The majority of computational modeling of tissue effects uses finite element models. These models can take the morphology of the cortex and the conductivities of various tissues into account to predict the spread of mechanical sound waves for ultrasound or of electrical currents from anode to cathode. Recent modeling studies include cortical folding and varying skull thickness to get personalized models of stimulation effects. While skull thickness is normally measured through computed tomography (CT), a technique that is not preferable for healthy subjects, there is ongoing development of MRI protocols that can replace CT.

Realistic head models are helpful for experimental design and accuracy in tissue targeting. There are several software systems provided by device manufacturers and also free software toolboxes. However, such models do not make predictions about the physiological effects, nor do they include the biological mechanisms that are affected by stimulation.

Modeling the Effect of Stimulation on Brain Tissue

The next step is to determine the effect of brain stimulation at the target site, and computational models can be applied to predict and understand what effect stimulation has on the target tissue (Wang et al., 2015). There are different levels of detail: (1) the neuronal description level, where the firing and sometimes dendritic activity of single neurons are modeled; and (2) the population description level, where the activity of an entire neural population is described, often by population average variables (see figure 15.2). Both approaches are used in the modeling of brain stimulation, each with their advantages and disadvantages. For example, the neuronal description level is easier to relate to measurable biological entities, but the amount of detail (and hence parameters) can be overwhelming and difficult to understand and to analyze. The population description level is often easier to understand mechanistically, with a limited number of parameters. However, the parameters are often more difficult to interpret or measure, and cellular processes are often included only in an abstract fashion. The choice of the description level is often dependent on the specific research question, the data available to validate the model against, and the need for mechanistic understanding.

There are several toolboxes to model the effects of brain stimulation: Lead-DBS to model deep-brain stimulation at the population level (Horn and Kuhn, 2015) and VERTEX to model the effect of electrical stimulation at the neuronal level (Thornton et al., 2019). In addition, there are several models for the effect of optogenetic stimulation concerning channelrhodopsin (Foutz et al., 2012; Nikolic et al., 2013).

Such models, following the framework of computational neurostimulation to study what brain stimulation does and how it affects behavior, can help to elucidate how differences between stimulation protocols and between individuals can alter the effect of stimulation. Using a detailed model of cortical tissue, including multiple layers and neuron types, could reproduce experimental effects of noninvasive electrical stimulation such as depolarization in individual pyramidal neurons, acceleration of intrinsic oscillations, and retention of the spatial profile of oscillations in different layers (Hutchings et al., 2020). However, the model also indicated that the extent of the effect critically depended on the angle at which the current is passing through the tissue: for some angles twice as much current was needed to yield the same effect on changing neuronal activity. This might explain why the same locations of tDCS electrodes across subjects could have a variable effect as the angle of current flow would vary with the cortical folding pattern of each subject.

Including synaptic plasticity in a detailed neuronal model of invasive focal stimulation in rat neocortex (Thornton et al., 2019) showed direct changes in neural activity and changes in synaptic plasticity: paired pulse stimulation induced short-term plasticity, and theta burst stimulation induced long-term potentiation. Such detailed models are useful when the experimental data for comparison includes local field potential recordings or information about layer organization and neuronal morphology. In particular, reproducing the layer architecture of different cortical and subcortical regions

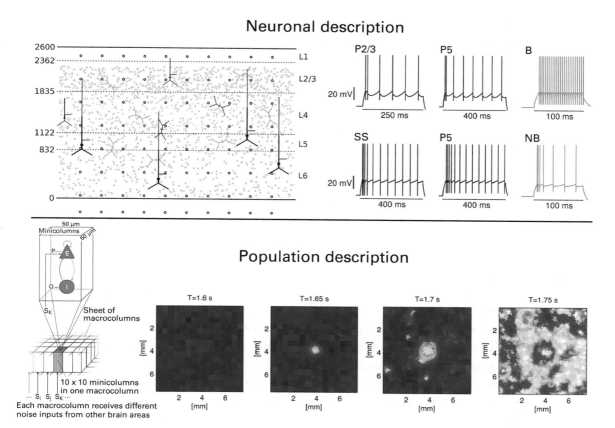

Figure 15.2
Different description levels of computer models of brain stimulation. Top: Neuronal description level, where the neurons in the different cortical layers are simulated. The multicompartment dendritic tree structure is shown for some of the neurons. The structure of an example cortical patch of tissue is shown on the left. Numbers on the left indicate cortical depth in micrometers. Some example time series of the firing dynamics of excitatory (black) and inhibitory (magenta) neurons are shown on the right. P = pyramidal cell, B = basket interneuron, NB = non-basket interneuron, SS = spiny stellate neuron, and numbers indicating cortical layer (after Tomsett et al., 2015). Bottom: Population-level description, where average excitatory and inhibitory population firing dynamics are simulated. The structure of the cortical slab consisting of multiple minicolumns is shown on the left. An example time series is shown of the dynamics at different time points (for details, see Wang et al., 2017).

could help us to understand how effects might differ depending on the stimulation target.

Finding Stimulation Targets

Up to now, we have only looked at the local effect of stimulation within a region, but there are also models of how stimulation changes the dynamics at the global scale. Many such models use control theory to identify the driver nodes that determine the dynamics of the entire network (Liu et al., 2011). The hope here is that stimulating a selected small number of nodes, ideally just one node, could be sufficient to alter the network dynamics in the desired direction. Therefore, they could be used to predict the most suitable stimulation targets.

Control theory can help us to understand the role of individual network nodes related to network dynamics. For *C. elegans*, predictions from control theory were able to indicate the roles of neurons regarding motor function, which was confirmed experimentally after the ablation of those neurons (Yan et al., 2017). Another application is to change the synchronization patterns of the network, in particular trying to prevent hypersynchronization. Using a numerical simulation of small-world or random networks of Kuramoto oscillators, stimulation based on control theory was compared with a standard Proportional-Differential Feedback control technique (Asllani et al., 2018). In both cases, the control theory–based approach was able to desynchronize the network while using less stimulation signal strength and/or fewer stimulation sites/electrodes than the standard control technique. Therefore, such models might reduce the number of stimulation targets and, crucially for implanted devices, the amount of energy that is used for altering network dynamics.

Models Applied to Different Brain Network Disorders

Brain stimulation has been applied to many brain disorders, and there is increasing interest in including connectome information when planning interventions (Luft et al., 2014). The following example studies use information about structural and functional connectivity to inform interventions in epilepsy, Parkinson's disease, and Alzheimer's disease.

For epilepsy, brain stimulation has been considered in order to prevent seizures or, at least, terminate seizures when they occur. However, there is the challenge of finding an optimal stimulation protocol for each individual patient. Including patient-specific structural connectivity, a model of epileptic spike-wave dynamics was used to simulate the effect of brain stimulation on seizure termination (Taylor, Thomas, et al., 2015). This model uses a pseudospectral method to determine the optimal control parameters in terms of where, when, and how to stimulate the network. Such nonlinear optimal control approaches could be useful for clinical interventions as the optimal stimulation configuration, informed by structural connectivity, varies between patients.

For Parkinson's disease, there are computational models of the neuronal circuits that deep-brain stimulation is targeting (Wiecki and Frank, 2010), featuring dopaminergic neurons innervating the striatum, dopamine involved in reinforcement learning, and models of the basal ganglia. Still, choosing the right stimulation targets and protocols for each individual patient is a challenge. A model that observes functional connectivity in patients was used to simulate the effects of stimulating different brain regions (Chen et al., 2018). In particular, the model observes to what extent abnormal topological features of functional connectivity for patients can be brought back closer to the features observed in healthy controls. The best predicted stimulation targets varied between patients and included the globus pallidus, the subthalamic nucleus, and, less frequently, other regions. Looking at clinical effects of stimulation in patients, stimulation-induced reductions of functional connectivity in these targets optimally maximized the therapeutic effects. Moreover, motor symptom severity correlated with the target rank of the subthalamic nucleus. Indeed, structural connectivity informed models (Horn and Kuhn, 2015) indicate that the subthalamic nucleus is involved in cognitive aspects of cautious motor preparation and kinematic gain control (Neumann et al., 2018). Altogether, this indicates that the change in functional connectivity, getting closer to the topology of healthy controls, could be used to inform the targets for deep-brain stimulation.

For Alzheimer's disease, there is often high neuronal activity and excitability during the early stages of disease progression. A computational model using structural and functional connectivity for connections between regions and a population model with an excitatory and inhibitory population within regions was used to study the effects of network stimulation (de Haan et al., 2017). The baseline model, without stimulation, could reproduce the activity-dependent degeneration with changed functional connectivity as observed in patients. Within the computer model testing scenarios where all regions were stimulated, the most successful outcomes occurred for targeting the excitatory populations within each region. This counterintuitive result, as one would expect targeting inhibitory populations could reduce hyperexcitability, shows how useful computational models can be in assessing treatment scenarios. Finally, both early-stage and late-stage stimulation interventions led to reductions in hyperexcitability, which, if confirmed by clinical trials, could widen the scope of patients who can be treated, as early dementia diagnosis is a major challenge. Using structural connectivity in Alzheimer patients, representing each region as a Duffing oscillator, control theory also identified several potential target regions and subjects to successfully respond to brain stimulation (Sanchez-Rodriguez et al., 2018).

Brain stimulation is an increasingly popular therapeutic tool, but it is not understood why some subjects do not respond to the stimulation or show severe side effects (Cyron, 2016). A systematic, rational design of brain stimulation adapted to each patient, used along with other approaches to optimize treatment, is lacking. Furthermore, one should only deliver the stimulus when necessary (closed-loop approach) and only control abnormal activity. In short, highly targeted patient-specific closed-loop brain stimulation should be rationally designed based on mechanistic insight. This will enable brain stimulation to become a truly integrated and tolerated part of the brain and could help to address ethical concerns about the risk of adverse effects in patients or healthy individuals (Cohen Kadosh et al., 2012).

This book has outlined the study of connectome changes over time in health and disease. For cognitive features, the hope is that understanding how such functions arise during development can give additional insights into how such features are implemented in the brain. For brain disorders, understanding the developmental origin of the disease state of an individual patient could give additional insights into the most suitable interventions. Changing connectomes can mean observing how connectomes change but also actively altering their organization or developmental trajectory. Brain stimulation has the potential to fine-tune the wiring of the brain in the case of brain network disorders, but only if potential long-term effects and side effects can be assessed before performing an intervention. I hope that this book can be a starting point toward such applications.

Glossary

Adjacency (connection) matrix The adjacency matrix of a *graph* is an $n \times n$ matrix with entries $a_{ij} = 1$ if node j connects to node i, and $a_{ij} = 0$ if there is no connection from node i to node j. Sometimes, nonzero entries indicate the strength of a connection using ordinal scales for fiber strength or metric scales [–1; 1] for correlation networks.

Adjacency list List where each line represents one edge with information about the source node, the target node, and (optionally) the weight of the edge connecting both nodes.

Anatomical connectivity The set of physical or structural (synaptic) connections linking neuronal units at a given time. Anatomical connectivity data can range over multiple spatial scales, from local circuits to large-scale networks of interregional pathways. Anatomical connection patterns are relatively static at shorter time scales (seconds to minutes) but may be dynamic at longer time scales (hours to days), for example, during learning or development.

Average shortest path The average shortest path *ASP* is the global mean of the entries of the *distance matrix*. Normally, infinite values (nonexisting paths) and values across the diagonal (loops) are not taken into account for the calculation of the mean of the distance matrix.

Betweenness centrality What proportion of all *shortest paths* are going through a node or edge of the network. These values are then called node betweenness or edge betweenness, respectively. See Brandes and Erlebach (2005) for other measures of centrality or influence.

Characteristic path length The characteristic path length L (also called "path length" or "average shortest path") is given by the global mean of the finite entries of the distance matrix. In some cases, the median or the harmonic mean may provide better estimates.

Chemotaxis (Greek taxis—arrangement/order) Orientation within chemical gradients.

Clustering coefficient The clustering coefficient of a node i is calculated as the number of existing connections between the node's neighbors divided by all their possible connections. The clustering coefficient ranges between 0 and 1 and is typically averaged over all nodes of a graph to yield the graph's clustering coefficient C.

Component A component is a set of nodes, for which every pair of nodes is joined by at least one path.

Component placement optimization (CPO) See *Optimal component placement*.

Connectedness A connected graph has only one component. A disconnected graph has at least two components.

Critical period A maturational stage in the life span of an organism during which the nervous system is especially sensitive to certain environmental stimuli.

Cycle A *path* that links a node to itself.

Degree The degree of a node is the sum of its incoming (afferent) and outgoing (efferent) connections. The number of afferent and efferent connections is also called the in-degree and out-degree, respectively.

Degree distribution Probability distribution of the degrees of all nodes of the network.

Degree sequence The N-tuple $(k_1, \ldots, k_N)$ with k_i as degree of node i and $k_i \leq k_{i+1}$ is called degree sequence.

Density (edge density) Proportion of edges (or arcs) existing in the network to the number of all possible edges (arcs).

Disconnection syndrome Originally, effects of lesions of association pathways (Geschwind, 1965).

Dispersion A measure of whether fibers from one region target many other regions or only few distant other regions. Fasciculation reduces the spreading of fibers and leads to lower dispersion.

Distal (Latin *distare*—standing apart) Located further away from the center.

Distance The distance between a source node j and a target node i is equal to the length of the shortest path.

Distance matrix The entries of the distance matrix correspond to the distance between node j and i. If no path exists, the distance between both nodes is set to infinity.

Dysconnection (Greek *dys*—bad or ill) A connection that is either reduced or increased in patients compared to healthy controls (Stephan et al., 2006).

Effective connectivity Describes the set of causal effects of one neural system over another. Thus, unlike functional connectivity, effective connectivity is not "model-free" but requires the specification of a causal model including structural parameters. Experimentally, effective connectivity may be inferred through perturbations, or through the observation of the temporal ordering of neural events.

Eutelic A species where the number of cells in their nervous system does not vary between individuals, but where that number is fixed.

Exponential graph Erdös-Rényi random graph with binomial degree distribution that can be fitted by an exponential function. In this book, such a graph is referred to as *random network*.

Glossary

Fragmentation Process where a network breaks apart into multiple components where paths to other nodes within a component exist (see *Reachability*) but not between components.

Functional compensation Lesions may have smaller effect, in particular if remaining regions are able to take over the function. There are three types of functions after lesions. First, residual function describes performance that remains after lesions of the mature brain. Second, recovered function describes neural and behavioral performance that emerges from, and is superior to, the initial residual function. Usually "recovery" assumes a prior presence of a given function (however, present to a lower extent) and does not include pathway rewiring. Third, spared functions are functions remaining after early lesion in the developing brain resulting from either an unmasking of existing pathways or the altered development of brain pathways.

Functional connectivity Captures patterns of deviations from statistical independence between distributed and often spatially remote neuronal units, measuring their correlation/covariance, spectral coherence, or phase-locking. Functional connectivity is time-dependent (hundreds of milliseconds) and "model-free"; that means measuring statistical interdependence (mutual information) without explicit reference to causal effects.

Functional plasticity Change of connection weights due to Hebbian learning.

Ganglion (plural: ganglia) A spatially concentrated group of neuron cell bodies.

Giant component The largest component, having more nodes than any other network component.

Graph Graphs are a set of *n* nodes (vertices, points, units) and *k* edges (connections, arcs). Graphs may be undirected (all connections are symmetrical) or directed. Because of the polarized nature of most neural connections, I focus on directed graphs, also called digraphs. In addition, graphs are simple; that means multiple (undirected) edges between nodes or loops (connections of one node to itself) do not exist.

Gyrencephalic Brains where the brain surface contains folds (gyri) and grooves (sulci). Species with gyrencephalic brains include human, cat, and macaque monkey.

Hebbian learning Change of connection strength where simultaneous firing of neurons leads to a strengthening of connections while asynchronous firing leads to a weakening of connection strength or synaptic efficacy (Hebb, 1949).

Hodology The study of pathways. The word is used in several contexts. (1) In brain physiology, it is the study of the interconnections of brain cells. (2) In philosophy, it is the study of interconnected ideas. (3) In geography, it is the study of paths.

Homeostasis The retainment of electrophysiological response properties, despite a changing environment, through adapting ion channel and synapse distributions across the neuron.

Hub Node with a degree that is much higher than for other nodes of the network. In sparse networks, even the degree of a hub might be relatively low.

Isolated node Node with no connections.

Kronecker symbol δ $\delta_{i,j} = 1$ for $i = j$ and $\delta_{i,j} = 0$ otherwise.

Leaf node Node with only one connection.

Linear graph Graph with many linear chains of nodes which can be detected by the *clustering coefficient* being lower than the *density*.

Lissencephalic Brains with a smooth surface, without folds (gyri) and grooves (sulci). Species with lissencephalic brains include rodents (mouse and rat) and the marmoset monkey.

Local shortest path For a node, the average length of the shortest path from that node to all other nodes in the network.

Loop A connection of a node onto itself (in other words: a cycle of length 1).

Matching index The matching index of two network nodes is defined as the average percentage of identical incoming or outgoing directed edges of the two nodes.

Minimum spanning tree A minimum spanning tree is a tree that connects all nodes of a network with a weight less than or equal to the weight of every other spanning tree. In this context, the weight is usually the total metric wiring length of the tree when the network is a *spatial graph*.

Modular graph A network with multiple modules.

Motif Subgraph with a certain number of nodes (usually 3, 4, or 5) that occurs significantly more often in a given network than in rewired networks with the same degree distribution.

Network flow The maximum flow between two nodes; for binary networks, corresponds to the number of alternative paths between both nodes.

Neurite Collective term for all parts growing out of the soma of a neuron, that is, axons and dendrites.

Nonlinear growth The number of nodes that is added at each step is changing over time. *Exponential growth* where the number of new nodes increases over time can lead to hub and rich-club organization.

Nucleus (plural: nuclei) A cluster of neurons in the central nervous system, located deep within the cerebral hemispheres and brain stem. The neurons in one nucleus usually have roughly similar connections and functions.

Old-gets-richer Nodes that are established early on can receive connections from all later forming nodes, leading to a higher ratio of incoming connections and a higher total number of connections. Nodes that are formed late will receive fewer incoming connections as most nodes have already matured, leading to a lower overall number of connections.

Optimal component placement (also: Component placement optimization; CPO) Nodes of a spatial network are arranged in an optimal way; that means every permutation of node positions (with connections being unchanged) would lead to a higher total wiring length.

Glossary

Parallel growth Extreme case where all nodes start to form connections at the same time. This leads to more bidirectional connections and fewer long-distance connections.

Participation coefficient The participation coefficient of a node measures what proportion of its links connect to other modules, outside its own module: the coefficient is close to 1 if its links are uniformly distributed among all the modules and 0 if all its links are within its own module (Guimera and Amaral, 2005).

Path A path is an ordered sequence of distinct connections and nodes, linking a source node j to a target node i. No connection or node is visited twice in a given path.

Path length The length of a *path* is equal to the number of distinct connections.

Plasticity Change of connections. See *Structural plasticity, Functional plasticity, Hebbian learning*.

Random network A graph with uniform connection probabilities and a binomial degree distribution. All nodes have roughly the same degree ("single-scale").

Reachability A set of nodes is reachable if a path between all pairs of nodes exists. Nodes in separate compartments are unreachable. Note that not only a lack of connections but also a high proportion of one-way links, running in the "wrong" direction, can lead to unreachable sets of nodes.

Scale (here, spatial extent of the nervous system) At early stages of development, neurons are in close proximity to one another, enabling easier formation of connections. At later stages, after the nervous system has expanded, neurons will be further apart with early formed (short) connections transforming into long-distance connections at the adult stage.

Scale-free graph Graph with a power-law degree distribution. "Scale-free" means that degrees are not grouped around one characteristic average degree (scale) but can spread over a very wide range of values, spanning several orders of magnitude.

Self-organized criticality (SOC) Many systems of interconnected, nonlinear elements are at a critical state in which event sizes (e.g., number of active nodes of a network) are scale-free and can be characterized by a power law. Reaching such a critical state is an emergent property of the system.

Serial growth Extreme case where nodes form connections one after another (i.e., the next node in the sequence only starts axon growth after the previous node has finished axon growth). This leads to fewer bidirectional connections, more long-distance connections, and nodes that establish connections early on more likely becoming network hubs.

Small-world graph A graph where the *clustering coefficient* is much higher than in a comparable random network but the *characteristic path length* remains about the same. The term "small world" was coined in relation to the notion that any two persons can be linked over few intermediate acquaintances.

Spatial graph Graphs or networks that extent in space; that means that every node has a spatial position. Spatial graphs are usually analyzed as being two- or three-dimensional, but naturally more dimensions are possible.

Spatial growth   A new method to yield *spatial graphs*. Starting with one node, at each step a new node with a random spatial position is added to the network. Then, the probability of the new node establishing connections with existing nodes decays with the spatial distance (e.g., exponential decay). Thereby, connections to nearby nodes are more likely than to distant nodes. If a node does not establish any connections, it is removed from the network. The procedure is repeated until the desired number of nodes is generated.

Structural plasticity   Addition or removal of connections or synapses.

Time windows   Nodes that are formed at the same time often come from the same cell lineage, are spatially nearby, and inherit the same start and end points for axon and synapse formation. Due to spatial proximity and overlapping times for connection formation, nodes with the same time window form topological modules.

Tropism   The turning of dendritic branches in a particular direction in response to an external stimulus.

References

Achacoso TB, and Yamamoto WS. 1992. *AY's neuroanatomy of C. elegans for computation*. Boca Raton, FL: CRC Press.

Achard S, and Bullmore E. 2007. Efficiency and cost of economical brain functional networks. *PLOS Comput Biol* 3 (2): e17.

Achard S, Salvador R, Whitcher B, Suckling J, and Bullmore E. 2006. A resilient, low-frequency, small-world human brain functional network with highly connected association cortical hubs. *J Neurosci* 26 (1): 63–72.

Addington AM, and Rapoport JL. 2012. Annual research review: Impact of advances in genetics in understanding developmental psychopathology. *J Child Psychol Psychiatry* 53 (5): 510–518.

Aerts H, Fias W, Caeyenberghs K, and Marinazzo D. 2016. Brain networks under attack: Robustness properties and the impact of lesions. *Brain* 139 (Pt 12): 3063–3083.

Ahn YY, Bagrow JP, and Lehmann S. 2010. Link communities reveal multiscale complexity in networks. *Nature* 466 (7307): 761–764.

Akil H, Martone ME, and Van Essen DC. 2011. Challenges and opportunities in mining neuroscience data. *Science* 331 (6018): 708–712.

Alexander-Bloch AF, Gogtay N, Meunier D, Birn R, Clasen L, et al. 2010. Disrupted modularity and local connectivity of brain functional networks in childhood-onset schizophrenia. *Front Syst Neurosci* 4: 147.

Alexander-Bloch AF, Vertes PE, Stidd R, Lalonde F, Clasen L, et al. 2013. The anatomical distance of functional connections predicts brain network topology in health and schizophrenia. *Cereb Cortex* 23 (1): 127–138.

Alfaro-Almagro F, Jenkinson M, Bangerter NK, Andersson JLR, Griffanti L, et al. 2018. Image processing and quality control for the first 10,000 brain imaging datasets from UK Biobank. *Neuroimage* 166: 400–424.

Alstott J, Breakspear M, Hagmann P, Cammoun L, and Sporns O. 2009. Modeling the impact of lesions in the human brain. *PLOS Comput Biol* 5 (6): e1000408.

Amari S. 1977. Dynamics of pattern formation in lateral-inhibition type neural fields. *Biol Cybern* 27 (2): 77–87.

American Psychiatric Association. 2013. *Diagnostic and statistical manual of mental disorders: DSM-5*. Washington, DC: American Psychiatric Publishing.

Anagnostou E, and Taylor MJ. 2011. Review of neuroimaging in autism spectrum disorders: What have we learned and where we go from here. *Mol Autism* 2 (1): 4.

Andersen SL. 2003. Trajectories of brain development: Point of vulnerability or window of opportunity? *Neurosci Biobehav Rev* 27 (1–2): 3–18.

Aparício D, Ribeiro P, and Silva F. 2017. Extending the applicability of graphlets to directed networks. *IEEE ACM Trans Comp Biol and Bioinf* 14 (6): 1302–1315.

Armstrong E, Schleicher A, Omran H, Curtis M, and Zilles K. 1995. The ontogeny of human gyrification. *Cereb Cortex* 5 (1): 56–63.

Arnatkeviciute A, Fulcher BD, and Fornito A. 2019. Uncovering the transcriptional correlates of hub connectivity in neural networks. *Front Neural Circuits* 13: 47.

Ascoli GA, Donohue DE, and Halavi M. 2007. Neuromorpho.Org: A central resource for neuronal morphologies. *J Neurosci* 27 (35): 9247–9251.

Asllani M, Expert P, and Carletti T. 2018. A minimally invasive neurostimulation method for controlling abnormal synchronisation in the neuronal activity. *PLOS Comput Biol* 14 (7): e1006296.

Babiloni C, Del Percio C, Lizio R, Noce G, Lopez S, et al. 2018. Abnormalities of resting-state functional cortical connectivity in patients with dementia due to Alzheimer's and Lewy body diseases: An EEG study. *Neurobiol Aging* 65: 18–40.

Bakola S, Burman KJ, and Rosa MG. 2015. The cortical motor system of the marmoset monkey (*Callithrix jacchus*). *Neurosci Res* 93: 72–81.

Ball G, Beare R, and Seal ML. 2017. Network component analysis reveals developmental trajectories of structural connectivity and specific alterations in autism spectrum disorder. *Hum Brain Mapp* 38 (8): 4169–4184.

Barabasi AL, and Albert R. 1999. Emergence of scaling in random networks. *Science* 286 (5439): 509–512.

Barbeito-Andres J, Gleiser PM, Bernal V, Hallgrimsson B, and Gonzalez PN. 2018. Brain structural networks in mouse exposed to chronic maternal undernutrition. *Neuroscience* 380: 14–26.

Barone P, Dehay C, Berland M, and Kennedy H. 1996. Role of directed growth and target selection in the formation of cortical pathways: Prenatal development of the projection of area v2 to area v4 in the monkey. *J Comp Neurol* 374 (1): 1–20.

Baruch L, Itzkovitz S, Golan-Mashiach M, Shapiro E, and Segal E. 2008. Using expression profiles of *Caenorhabditis elegans* neurons to identify genes that mediate synaptic connectivity. *PLOS Comput Biol* 4 (7): e1000120.

References

Bassett DS, Bullmore E, Verchinski BA, Mattay VS, Weinberger DR, et al. 2008. Hierarchical organization of human cortical networks in health and schizophrenia. *J Neurosci* 28 (37): 9239–9248.

Bassett DS, Xia CH, and Satterthwaite TD. 2018. Understanding the emergence of neuropsychiatric disorders with network neuroscience. *Biol Psychiatry Cogn Neurosci Neuroimaging* 3 (9): 742–753.

Bathelt J, Gathercole SE, Butterfield S, the CALM team, and Astle DE. 2018. Children's academic attainment is linked to the global organization of the white matter connectome. *Dev Sci* 21 (5): e12662.

Bauer R, Clowry GJ, and Kaiser M. 2020. Creative destruction: A basic computational model of cortical layer formation. *bioRxiv*, doi:10.1101/2020.01.29.921999.

Bauer R, and Kaiser M. 2017. Nonlinear growth: An origin of hub organization in complex networks. *R Soc Open Sci* 4 (3): 160691.

Bauer R, Kaiser M, and Stoll E. 2014. A computational model incorporating neural stem cell dynamics reproduces glioma incidence across the lifespan in the human population. *PLOS One* 9 (11): e111219.

Bauer R, Zubler F, Pfister S, Hauri A, Pfeiffer M, et al. 2014. Developmental self-construction and -configuration of functional neocortical neuronal networks. *PLOS Comput Biol* 10 (12): e1003994.

Bayer SA, and Altman J. 1991. *Neocortical development*. New York: Raven Press.

Bayly PV, Okamoto RJ, Xu G, Shi Y, and Taber LA. 2013. A cortical folding model incorporating stress-dependent growth explains gyral wavelengths and stress patterns in the developing brain. *Phys Biol* 10 (1): 016005.

Belfore LA, Johnson B, and Aylor J. 1989. "Modeling of fault tolerance in neural networks." 15th Annual Conference of IEEE Industrial Electronics Society.

Belmonte MK, Allen G, Beckel-Mitchener A, Boulanger LM, Carper RA, et al. 2004. Autism and abnormal development of brain connectivity. *J Neurosci* 24 (42): 9228–9231.

Bengtsson SL, Nagy Z, Skare S, Forsman L, Forssberg H, et al. 2005. Extensive piano practicing has regionally specific effects on white matter development. *Nat Neurosci* 8 (9): 1148–1150.

Bernardoni F, King JA, Geisler D, Birkenstock J, Tam FI, et al. 2018. Nutritional status affects cortical folding: Lessons learned from anorexia nervosa. *Biol Psychiatry* 84 (9): 692–701.

Bestmann S, de Berker AO, and Bonaiuto J. 2015. Understanding the behavioural consequences of noninvasive brain stimulation. *Trends Cogn Sci* 19 (1): 13–20.

Betzel RF, Avena-Koenigsberger A, Goni J, He Y, de Reus MA, et al. 2016. Generative models of the human connectome. *Neuroimage* 124 (Pt A): 1054–1064.

Betzel RF, Byrge L, He Y, Goni J, Zuo XN, et al. 2014. Changes in structural and functional connectivity among resting-state networks across the human lifespan. *Neuroimage* 102 (Pt 2): 345–357.

Beul SF, Grant S, and Hilgetag CC. 2015. A predictive model of the cat cortical connectome based on cytoarchitecture and distance. *Brain Struct Funct* 220 (6): 3167–3184.

Bhagwat N, Viviano JD, Voineskos AN, Chakravarty MM, and Alzheimer's Disease Neuroimaging Initiative. 2018. Modeling and prediction of clinical symptom trajectories in Alzheimer's disease using longitudinal data. *PLOS Comput Biol* 14 (9): e1006376.

Bienkowski MS, Bowman I, Song MY, Gou L, Ard T, et al. 2018. Integration of gene expression and brain-wide connectivity reveals the multiscale organization of mouse hippocampal networks. *Nat Neurosci* 21 (11): 1628–1643.

Binzegger T, Douglas RJ, and Martin KA. 2004. A quantitative map of the circuit of cat primary visual cortex. *J Neurosci* 24 (39): 8441–8453.

Binzegger T, Douglas RJ, and Martin KA. 2005. Axons in cat visual cortex are topologically self-similar. *Cereb Cortex* 15 (2): 152–165.

Binzegger T, Douglas RJ, and Martin KA. 2009. Topology and dynamics of the canonical circuit of cat V1. *Neural Netw* 22 (8): 1071–1078.

Blaxter M. 2011. Nematodes: The worm and its relatives. *PLOS Biol* 9 (4): e1001050.

Boksa P. 2012. Abnormal synaptic pruning in schizophrenia: Urban myth or reality? *J Psychiatry Neurosci* 37 (2): 75–77.

Bollobas B. 1985. *Random graphs*. Cambridge: Cambridge University Press.

Bonifazi P, Goldin M, Picardo MA, Jorquera I, Cattani A, et al. 2009. GABAergic hub neurons orchestrate synchrony in developing hippocampal networks. *Science* 326 (5958): 1419–1424.

Bonilha L, Nesland T, Martz GU, Joseph JE, Spampinato MV, et al. 2012. Medial temporal lobe epilepsy is associated with neuronal fibre loss and paradoxical increase in structural connectivity of limbic structures. *J Neurol Neurosurg Psychiatry* 83 (9): 903–909.

Borisyuk R, Cooke T, and Roberts A. 2008. Stochasticity and functionality of neural systems: Mathematical modelling of axon growth in the spinal cord of tadpole. *Biosystems* 93 (1–2): 101–114.

Bota M, Dong HW, and Swanson LW. 2012. Combining collation and annotation efforts toward completion of the rat and mouse connectomes in BAMS. *Front Neuroinform* 6: 2.

Bozzi Y, Casarosa S, and Caleo M. 2012. Epilepsy as a neurodevelopmental disorder. *Front Psychiatry* 3: 19.

Braak H, Del Tredici K, Rub U, de Vos RA, Jansen Steur EN, et al. 2003. Staging of brain pathology related to sporadic Parkinson's disease. *Neurobiol Aging* 24: 197–211.

Braitenberg V, and Schüz A. 1998. *Cortex: Statistics and geometry of neuronal connectivity*. Heidelberg: Springer.

References

Brandes U, and Erlebach T. 2005. *Network analysis: Lecture notes in computer science.* Heidelberg: Springer.

Breakspear M, and Stam CJ. 2005. Dynamics of a neural system with a multiscale architecture. *Philos Trans R Soc Lond B Biol Sci* 360 (1457): 1051–1074.

Briscoe SD, and Ragsdale CW. 2018. Homology, neocortex, and the evolution of developmental mechanisms. *Science* 362 (6411): 190–193.

Bullmore E, and Sporns O. 2009. Complex brain networks: Graph theoretical analysis of structural and functional systems. *Nat Rev Neurosci* 10 (3): 186–198.

Bullmore ET. 2018. *The inflamed mind.* London: Short Books.

Bundy DT, and Nudo RJ. 2019. Preclinical studies of neuroplasticity following experimental brain injury. *Stroke* 50 (9): 2626–2633.

Burns GA, and Young MP. 2000. Analysis of the connectional organization of neural systems associated with the hippocampus in rats. *Philos Trans R Soc Lond B Biol Sci* 355 (1393): 55–70.

Butz M, Steenbuck ID, and van Ooyen A. 2014. Homeostatic structural plasticity increases the efficiency of small-world networks. *Front Synaptic Neurosci* 6: 7.

Butz M, Teuchert-Noodt G, Grafen K, and van Ooyen A. 2008. Inverse relationship between adult hippocampal cell proliferation and synaptic rewiring in the dentate gyrus. *Hippocampus* 18 (9): 879–898.

Butz M, and van Ooyen A. 2013. A simple rule for dendritic spine and axonal bouton formation can account for cortical reorganization after focal retinal lesions. *PLOS Comput Biol* 9 (10): e1003259.

Butz M, van Ooyen A, and Wörgötter F. 2009. A model for cortical rewiring following deafferentation and focal stroke. *Front Comput Neurosci* 3: 10.

Butz M, Wörgötter F, and van Ooyen A. 2009. Activity-dependent structural plasticity. *Brain Res Rev* 60 (2): 287–305.

Cabral J, Hugues E, Kringelbach ML, and Deco G. 2012. Modeling the outcome of structural disconnection on resting-state functional connectivity. *Neuroimage* 62 (3): 1342–1353.

Caffrey JR, Hughes BD, Britto JM, and Landman KA. 2014. An in silico agent-based model demonstrates Reelin function in directing lamination of neurons during cortical development. *PLOS One* 9 (10): e110415.

Cahalane DJ, Charvet CJ, and Finlay BL. 2014. Modeling local and cross-species neuron number variations in the cerebral cortex as arising from a common mechanism. *Proc Natl Acad Sci USA* 111 (49): 17642–17647.

Caleo M. 2009. Epilepsy: Synapses stuck in childhood. *Nat Med* 15 (10): 1126–1127.

Cao M, Huang H, and He Y. 2017. Developmental connectomics from infancy through early childhood. *Trends Neurosci* 40 (8): 494–506.

Castellanos NP, Paul N, Ordonez VE, Demuynck O, Bajo R, et al. 2010. Reorganization of functional connectivity as a correlate of cognitive recovery in acquired brain injury. *Brain* 133 (Pt 8): 2365–2381.

Catani M, and ffytche DH. 2005. The rises and falls of disconnection syndromes. *Brain* 128 (Pt 10): 2224–2239.

Chalmers K, Kita EM, Scott EK, and Goodhill GJ. 2016. Quantitative analysis of axonal branch dynamics in the developing nervous system. *PLOS Comput Biol* 12 (3): e1004813.

Chaudhuri D, Borowski P, and Zapotocky M. 2011. Model of fasciculation and sorting in mixed populations of axons. *Phys Rev E Stat Nonlin Soft Matter Phys* 84 (2 Pt 1): 021908.

Chavez M, Valencia M, Navarro V, Latora V, and Martinerie J. 2010. Functional modularity of background activities in normal and epileptic brain networks. *Phys Rev Lett* 104 (11): 118701.

Chechik G, Meilijson I, and Ruppin E. 1999. Neuronal regulation: A mechanism for synaptic pruning during brain maturation. *Neural Comput* 11 (8): 2061–2080.

Chen BL, Hall DH, and Chklovskii DB. 2006. Wiring optimization can relate neuronal structure and function. *Proc Natl Acad Sci USA* 103 (12): 4723–4728.

Chen H, Zhang T, Guo L, Li K, Yu X, et al. 2013. Coevolution of gyral folding and structural connection patterns in primate brains. *Cereb Cortex* 23 (5): 1208–1217.

Chen X, Zhang C, Li Y, Huang P, Lv Q, et al. 2018. Functional connectivity-based modelling simulates subject-specific network spreading effects of focal brain stimulation. *Neurosci Bull* 34 (6): 921–938.

Cherniak C. 1992. Local optimization of neuron arbors. *Biol Cybern* 66 (6): 503–510.

Cherniak C. 1994. Component placement optimization in the brain. *J Neurosci* 14 (4): 2418–2427.

Chiang AS, Lin CY, Chuang CC, Chang HM, Hsieh CH, et al. 2011. Three-dimensional reconstruction of brain-wide wiring networks in *Drosophila* at single-cell resolution. *Curr Biol* 21 (1): 1–11.

Chklovskii DB, Schikorski T, and Stevens CF. 2002. Wiring optimization in cortical circuits. *Neuron* 34 (3): 341–347.

Choe Y, McCormcik BH, and Koh W. 2004. Network connectivity analysis on the temporally augmented *C. elegans* web: A pilot study. *Soc Neurosci Abstr* 30: 921.929.

Clauset A, Moore C, and Newman ME. 2008. Hierarchical structure and the prediction of missing links in networks. *Nature* 453 (7191): 98–101.

Clauset A, Shalizi CR, and Newman MEJ. 2009. Power-law distributions in empirical data. *SIAM Review* 51 (4): 661–703.

References

Cohen Kadosh R, Levy N, O'Shea J, Shea N, and Savulescu J. 2012. The neuroethics of non-invasive brain stimulation. *Curr Biol* 22 (4): R108–111.

Colizza V, Flammini A, Serrano MA, and Vespignani A. 2006. Detecting rich-club ordering in complex networks. *Nature Phys* 2 (2): 110–115.

Collin G, Scholtens LH, Kahn RS, Hillegers MHJ, and van den Heuvel MP. 2017. Affected anatomical rich club and structural-functional coupling in young offspring of schizophrenia and bipolar disorder patients. *Biol Psychiatry* 82 (10): 746–755.

Collin G, and van den Heuvel MP. 2013. The ontogeny of the human connectome: Development and dynamic changes of brain connectivity across the life span. *Neuroscientist* 19 (6): 616–628.

Collins CE. 2011. Variability in neuron densities across the cortical sheet in primates. *Brain, Behavior and Evolution* 78 (1): 37–50.

Contreras JA, Avena-Koenigsberger A, Risacher SL, West JD, Tallman E, et al. 2019. Resting state network modularity along the prodromal late onset Alzheimer's disease continuum. *Neuroimage Clin* 22: 101687.

Cook SJ, Jarrell TA, Brittin CA, Wang Y, Bloniarz AE, et al. 2019. Whole-animal connectomes of both *Caenorhabditis elegans* sexes. *Nature* 571 (7763): 63–71.

Cope TE, Rittman T, Borchert RJ, Jones PS, Vatansever D, et al. 2018. Tau burden and the functional connectome in Alzheimer's disease and progressive supranuclear palsy. *Brain* 141 (2): 550–567.

Cormen TH, Leiserson CE, Rivest RL, and Stein C. 2009. *Introduction to algorithms*. 3rd ed. Cambridge: MIT Press.

Costa L da F, Kaiser M, and Hilgetag CC. 2007. Predicting the connectivity of primate cortical networks from topological and spatial node properties. *BMC Syst Biol* 1: 16.

Costa L da F, Rodrigues FA, Travieso G, and Boas PRV. 2007. Characterization of complex networks: A survey of measurements. *Adv Phys* 56 (1): 167–242.

Costa L da F, Zawadzki K, Miazaki M, Viana MP, and Taraskin SN. 2010. Unveiling the neuromorphological space. *Front Comput Neurosci* 4: 150.

Coupe P, Manjon JV, Lanuza E, and Catheline G. 2019. Lifespan changes of the human brain in Alzheimer's disease. *Sci Rep* 9 (1): 3998.

Courchesne E, and Pierce K. 2005. Why the frontal cortex in autism might be talking only to itself: Local over-connectivity but long-distance disconnection. *Curr Opin Neurobiol* 15 (2): 225–230.

Cowan WM, Fawcett JW, O'Leary DD, and Stanfield BB. 1984. Regressive events in neurogenesis. *Science* 225 (4668): 1258–1265.

Crawley JN. 2012. Translational animal models of autism and neurodevelopmental disorders. *Dialogues in clinical neuroscience* 14 (3): 293–305.

Crepel F. 1982. Regression of functional synapses in the immature mammalian cerebellum. *Trends Neurosci* 5 (8): 266–269.

Crick F, and Jones E. 1993. Backwardness of human neuroanatomy. *Nature* 361 (6408): 109–110.

Csernansky JG, Gillespie SK, Dierker DL, Anticevic A, Wang L, et al. 2008. Symmetric abnormalities in sulcal patterning in schizophrenia. *Neuroimage* 43 (3): 440–446.

Cyron D. 2016. Mental side effects of deep brain stimulation (DBS) for movement disorders: The futility of denial. *Front Integr Neurosci* 10: 17.

Damicelli F, Hilgetag CC, Hütt M-T, and Messé A. 2019. Topological reinforcement as a principle of modularity emergence in brain networks. *Network Neuroscience* 3 (2): 589–605.

Dancause N, Barbay S, Frost SB, Plautz EJ, Chen D, et al. 2005. Extensive cortical rewiring after brain injury. *J Neurosci* 25 (44): 10167–10179.

Dayan P, and Abbott LF. 2001. *Theoretical neuroscience: Computational and mathematical modeling of neural systems.* Cambridge, MA: MIT Press.

de Haan W, Mott K, van Straaten ECW, Scheltens P, and Stam CJ. 2012. Activity dependent degeneration explains hub vulnerability in Alzheimer's disease. *PLOS Computational Biology* 8 (8): e1002582.

de Haan W, van Straaten ECW, Gouw AA, and Stam CJ. 2017. Altering neuronal excitability to preserve network connectivity in a computational model of Alzheimer's disease. *PLOS Computational Biology* 13 (9): e1005707.

de Reus MA, and van den Heuvel MP. 2013. Rich club organization and intermodule communication in the cat connectome. *J Neurosci* 33 (32): 12929–12939.

Deco G, Jirsa VK, and McIntosh AR. 2011. Emerging concepts for the dynamical organization of resting-state activity in the brain. *Nat Rev Neurosci* 12 (1): 43–56.

DeFelipe J. 2010. From the connectome to the synaptome: An epic love story. *Science* 330 (6008): 1198–1201.

Deguchi Y, Donato F, Galimberti I, Cabuy E, and Caroni P. 2011. Temporally matched subpopulations of selectively interconnected principal neurons in the hippocampus. *Nat Neurosci* 14 (4): 495–504.

Dehaene S, Cohen L, Morais J, and Kolinsky R. 2015. Illiterate to literate: Behavioural and cerebral changes induced by reading acquisition. *Nat Rev Neurosci* 16 (4): 234–244.

Delli Pizzi S, Franciotti R, Taylor JP, Esposito R, Tartaro A, et al. 2015. Structural connectivity is differently altered in dementia with Lewy body and Alzheimer's disease. *Front Aging Neurosci* 7: 208.

References

Demirtas M, Tornador C, Falcon C, Lopez-Sola M, Hernandez-Ribas R, et al. 2016. Dynamic functional connectivity reveals altered variability in functional connectivity among patients with major depressive disorder. *Hum Brain Mapp* 37 (8): 2918–2930.

Denk W, and Horstmann H. 2004. Serial block-face scanning electron microscopy to reconstruct three-dimensional tissue nanostructure. *PLOS Biol* 2 (11): e329.

DeSalvo MN, Douw L, Tanaka N, Reinsberger C, and Stufflebeam SM. 2014. Altered structural connectome in temporal lobe epilepsy. *Radiology* 270 (3): 842–848.

Di Martino A, Fair DA, Kelly C, Satterthwaite TD, Castellanos FX, et al. 2014. Unraveling the miswired connectome: A developmental perspective. *Neuron* 83 (6): 1335–1353.

Diestel R. 1997. *Graph theory*. New York: Springer.

Douglas RJ, and Martin KA. 2004. Neuronal circuits of the neocortex. *Annu Rev Neurosci* 27: 419–451.

Douglas RJ, Martin KA, and Whitteridge D. 1989. A canonical microcircuit for neocortex. *Neural Computation* 1 (4): 480–488.

Druckmann S, Feng L, Lee B, Yook C, Zhao T, et al. 2014. Structured synaptic connectivity between hippocampal regions. *Neuron* 81 (3): 629–640.

Dubois B, Feldman HH, Jacova C, Hampel H, Molinuevo JL, et al. 2014. Advancing research diagnostic criteria for Alzheimer's disease: The IWG-2 criteria. *Lancet Neurol* 13 (6): 614–629.

Dumas de la Roque A, Oppenheim C, Chassoux F, Rodrigo S, Beuvon F, et al. 2005. Diffusion tensor imaging of partial intractable epilepsy. *Eur Radiol* 15 (2): 279–285.

Duncan J, Chylinski D, Mitchell DJ, and Bhandari A. 2017. Complexity and compositionality in fluid intelligence. *Proc Natl Acad Sci USA* 114 (20): 5295–5299.

Durbin RM. 1987. "Studies on the development and organization of the nervous system of *Caenorhabditis elegans*." PhD Thesis, King's College, Cambridge University.

Ebbesson SOE. 1980. The parcellation theory and its relation to interspecific variability in brain organization, evolutionary and ontogenetic development, and neuronal plasticity. *Cell Tissue Res* 213 (2): 179–212.

Ebbesson SOE. 1984. An update of the parcellation theory. *Behav Brain Sci* 7 (3): 350–360.

Eglen SJ, van Ooyen A, and Willshaw DJ. 2000. Lateral cell movement driven by dendritic interactions is sufficient to form retinal mosaics. *Network* 11 (1): 103–118.

Eglen SJ, and Willshaw DJ. 2002. Influence of cell fate mechanisms upon retinal mosaic formation: A modelling study. *Development* 129 (23): 5399–5408.

Eguiluz VM, Chialvo DR, Cecchi GA, Baliki M, and Apkarian AV. 2005. Scale-free brain functional networks. *Phys Rev Lett* 94 (1): 018102.

Eickhoff SB, Constable RT, and Yeo BTT. 2018. Topographic organization of the cerebral cortex and brain cartography. *Neuroimage* 170: 332–347.

Eickhoff SB, Yeo BTT, and Genon S. 2018. Imaging-based parcellations of the human brain. *Nat Rev Neurosci* 19 (11): 672–686.

Ercsey-Ravasz M, Markov NT, Lamy C, Van Essen DC, Knoblauch K, et al. 2013. A predictive network model of cerebral cortical connectivity based on a distance rule. *Neuron* 80 (1): 184–197.

Erdős P, and Rényi A. 1959. On random graphs, I. *Publicationes Mathematicae (Debrecen)* 6: 290–297.

Erdős P, and Rényi A. 1960. On the evolution of random graphs. *Publ Math Inst Hung Acad Sci* 5: 17–61.

Fair DA, Cohen AL, Power JD, Dosenbach NU, Church JA, et al. 2009. Functional brain networks develop from a "local to distributed" organization. *PLOS Comput Biol* 5 (5): e1000381.

Fard PK, Pfeiffer M, and Bauer R. 2018. A generative growth model for thalamocortical axonal branching in primary visual cortex. *bioRxiv*, doi:10.1101/288522.

Fauth MJ, and van Rossum MC. 2019. Self-organized reactivation maintains and reinforces memories despite synaptic turnover. *Elife* 8.

Felleman DJ, and Van Essen DC. 1991. Distributed hierarchical processing in the primate cerebral cortex. *Cereb Cortex* 1 (1): 1–47.

Fields RD, Woo DH, and Basser PJ. 2015. Glial regulation of the neuronal connectome through local and long-distant communication. *Neuron* 86 (2): 374–386.

Finger S, and Stein DG. 1982. *Brain damage and recovery*. New York: Academic Press.

Fitzgibbon SP, Jenkinson M, Robinson E, Bozek J, Griffanti L, et al. 2016. "The developing Human Connectome Project (dHCP): Minimal functional preprocessing pipeline for neonates." Fifth Biennial Conference on Resting State and Brain Connectivity.

Fornito A, Arnatkeviciute A, and Fulcher BD. 2019. Bridging the gap between connectome and transcriptome. *Trends Cogn Sci* 23 (1): 34–50.

Fornito A, Zalesky A, and Breakspear M. 2015. The connectomics of brain disorders. *Nat Rev Neurosci* 16 (3): 159–172.

Fornito A, Zalesky A, Pantelis C, and Bullmore ET. 2012. Schizophrenia, neuroimaging and connectomics. *Neuroimage* 62 (4): 2296–2314.

Foutz TJ, Arlow RL, and McIntyre CC. 2012. Theoretical principles underlying optical stimulation of a channelrhodopsin-2 positive pyramidal neuron. *J Neurophysiol* 107 (12): 3235–3245.

Franze K. 2013. The mechanical control of nervous system development. *Development* 140 (15): 3069–3077.

Fratiglioni L, Paillard-Borg S, and Winblad B. 2004. An active and socially integrated lifestyle in late life might protect against dementia. *Lancet Neurol* 3 (6): 343–353.

Fries P. 2015. Rhythms for cognition: Communication through coherence. *Neuron* 88 (1): 220–235.

Friston KJ. 1998. The disconnection hypothesis. *Schizophr Res* 30 (2): 115–125.

Gaig C, and Tolosa E. 2009. When does Parkinson's disease begin? *Mov Disord* 24 Suppl 2: S656–664.

Gallos LK, Makse HA, and Sigman M. 2012. A small world of weak ties provides optimal global integration of self-similar modules in functional brain networks. *Proc Natl Acad Sci USA* 109 (8): 2825–2830.

Galvan CD, Hrachovy RA, Smith KL, and Swann JW. 2000. Blockade of neuronal activity during hippocampal development produces a chronic focal epilepsy in the rat. *J Neurosci* 20 (8): 2904–2916.

Gao W, Gilmore JH, Giovanello KS, Smith JK, Shen D, et al. 2011. Temporal and spatial evolution of brain network topology during the first two years of life. *PLOS One* 6 (9): e25278.

Garcia KE, Kroenke CD, and Bayly PV. 2018. Mechanics of cortical folding: Stress, growth and stability. *Philos Trans R Soc Lond B Biol Sci* 373 (1759).

Garcia-Cabezas MA, Zikopoulos B, and Barbas H. 2019. The structural model: A theory linking connections, plasticity, pathology, development and evolution of the cerebral cortex. *Brain Struct Funct* 224 (3): 985–1008.

Georgiadis K, Wray S, Ourselin S, Warren JD, and Modat M. 2018. Computational modelling of pathogenic protein spread in neurodegenerative diseases. *PLOS One* 13 (2): e0192518.

Geschwind N. 1965. Disconnection syndromes in animals and man: Part I. *Brain* 88: 229–237.

Geser F, Wenning GK, Poewe W, and McKeith I. 2005. How to diagnose dementia with Lewy bodies: State of the art. *Mov Disord* 20 Suppl 12: S11–20.

Gewaltig M-O, and Diesmann M. 2007. NEST (NEural Simulation Tool). *Scholarpedia* 2 (4): 1430.

Gibson EM, Purger D, Mount CW, Goldstein AK, Lin GL, et al. 2014. Neuronal activity promotes oligodendrogenesis and adaptive myelination in the mammalian brain. *Science* 344 (6183): 1252304.

Giedd JN, Blumenthal J, Jeffries NO, Castellanos FX, Liu H, et al. 1999. Brain development during childhood and adolescence: A longitudinal MRI study. *Nat Neurosci* 2 (10): 861–863.

Gierer A. 1988. Spatial organization and genetic information in brain development. *Biol Cybern* 59: 13–21.

Gierer A, and Meinhardt H. 1972. A theory of biological pattern formation. *Kybernetik* 12: 30–39.

Gilbert CD, and Wiesel TN. 1989. Columnar specificity of intrinsic horizontal and corticocortical connections in cat visual cortex. *J Neurosci* 9 (7): 2432–2442.

Girvan M, and Newman ME. 2002. Community structure in social and biological networks. *Proc Natl Acad Sci USA* 99 (12): 7821–7826.

Glasser MF, Coalson TS, Robinson EC, Hacker CD, Harwell J, et al. 2016. A multi-modal parcellation of human cerebral cortex. *Nature* 536 (7615): 171–178.

Glasser MF, Sotiropoulos SN, Wilson JA, Coalson TS, Fischl B, et al. 2013. The minimal preprocessing pipelines for the Human Connectome Project. *Neuroimage* 80: 105–124.

Gleissner U, Sassen R, Schramm J, Elger CE, and Helmstaedter C. 2005. Greater functional recovery after temporal lobe epilepsy surgery in children. *Brain* 128 (Pt 12): 2822–2829.

Godfrey KB, Eglen SJ, and Swindale NV. 2009. A multi-component model of the developing retinocollicular pathway incorporating axonal and synaptic growth. *PLOS Comput Biol* 5 (12): e1000600.

Gohlke JM, Griffith WC, and Faustman EM. 2007. Computational models of neocortical neuronogenesis and programmed cell death in the developing mouse, monkey, and human. *Cereb Cortex* 17 (10): 2433–2442.

Goldsworthy MR, Pitcher JB, and Ridding MC. 2015. Spaced noninvasive brain stimulation: Prospects for inducing long-lasting human cortical plasticity. *Neurorehabil Neural Repair* 29 (8): 714–721.

Goodall S, Reggia JA, Chen Y, Ruppin E, and Whitney C. 1997. A computational model of acute focal cortical lesions. *Stroke* 28 (1): 101–109.

Goodfellow M, Rummel C, Abela E, Richardson MP, Schindler K, et al. 2016. Estimation of brain network ictogenicity predicts outcome from epilepsy surgery. *Sci Rep* 6: 29215.

Goodhill GJ. 1997. Diffusion in axon guidance. *Eur J Neurosci* 9 (7): 1414–1421.

Goodhill GJ, and Urbach JS. 1999. Theoretical analysis of gradient detection by growth cones. *J Neurobiol* 41 (2): 230–241.

Goslin K, and Banker G. 1989. Experimental observations on the development of polarity by hippocampal neurons in culture. *J Cell Biol* 108 (4): 1507–1516.

Goulas A, Betzel RF, and Hilgetag CC. 2019. Spatiotemporal ontogeny of brain wiring. *Science Advances* 5 (6): eaav9694.

Goulas A, Schaefer A, and Margulies DS. 2015. The strength of weak connections in the macaque cortico-cortical network. *Brain Struct Funct* 220 (5): 2939–2951.

Gouzé JL, Lasry JM, and Changeux JP. 1983. Selective stabilization of muscle innervation during development: A mathematical model. *Biol Cybern* 46: 207–215.

Granovetter MS. 1973. The strength of weak ties. *Am J Sociol* 78 (6): 1360–1380.

Grober E, and Buschke H. 1987. Genuine memory deficits in dementia. *Dev Neuropsychol* 3 (1): 13–36.

Grossman N, Bono D, Dedic N, Kodandaramaiah SB, Rudenko A, et al. 2017. Noninvasive deep brain stimulation via temporally interfering electric fields. *Cell* 169 (6): 1029–1041.e1016.

Guido W, Spear PD, and Tong L. 1990. Functional compensation in the lateral suprasylvian visual area following bilateral visual cortex damage in kittens. *Exp Brain Res* 83 (1): 219–224.

Guimera R, and Amaral LAN. 2005. Functional cartography of complex metabolic networks. *Nature* 433 (7028): 895–900.

Güntürkün O. 2005. The avian 'prefrontal cortex' and cognition. *Curr Opin Neurobiol* 15: 686–693.

Guy J, and Staiger JF. 2017. The functioning of a cortex without layers. *Front Neuroanat* 11: 54.

Hagmann P, Cammoun L, Gigandet X, Meuli R, Honey CJ, et al. 2008. Mapping the structural core of human cerebral cortex. *PLOS Biol* 6 (7): e159.

Hagmann P, Sporns O, Madan N, Cammoun L, Pienaar R, et al. 2010. White matter maturation reshapes structural connectivity in the late developing human brain. *Proc Natl Acad Sci USA* 107 (44): 19067–19072.

Hahn JD, Sporns O, Watts AG, and Swanson LW. 2019. Macroscale intrinsic network architecture of the hypothalamus. *Proc Natl Acad Sci USA* 116 (16): 8018–8027.

Haier RJ, Jung RE, Yeo RA, Head K, and Alkire MT. 2004. Structural brain variation and general intelligence. *Neuroimage* 23 (1): 425–433.

Hall DH, and Altun ZF. 2008. *C. elegans atlas*. Cold Spring Harbor, NY: Cold Spring Harbor Laboratory Press.

Hardan AY, Jou RJ, Keshavan MS, Varma R, and Minshew NJ. 2004. Increased frontal cortical folding in autism: A preliminary MRI study. *Psychiatry Res* 131 (3): 263–268.

Harris JA, Mihalas S, Hirokawa KE, Whitesell JD, Choi H, et al. 2019. Hierarchical organization of cortical and thalamic connectivity. *Nature*: 1–8.

Harris JM, Whalley H, Yates S, Miller P, Johnstone EC, et al. 2004. Abnormal cortical folding in high-risk individuals: A predictor of the development of schizophrenia? *Biol Psychiatry* 56 (3): 182–189.

Hebb DO. 1949. *The organization of behavior*. New York: John Wiley & Sons.

Hedderich DM, Bauml JG, Berndt MT, Menegaux A, Scheef L, et al. 2019. Aberrant gyrification contributes to the link between gestational age and adult IQ after premature birth. *Brain* 142 (5): 1255–1269.

Helfrich RF, Knepper H, Nolte G, Struber D, Rach S, et al. 2014. Selective modulation of interhemispheric functional connectivity by HD-tACS shapes perception. *PLOS Biol* 12 (12): e1002031.

Hellwig B. 2000. A quantitative analysis of the local connectivity between pyramidal neurons in layers 2/3 of the rat visual cortex. *Biol Cybern* 82 (2): 111–121.

Hellwig B, Schüz A, and Aertsen A. 1994. Synapses on axon collaterals of pyramidal cells are spaced at random intervals: A Golgi study in the mouse cerebral cortex. *Biol Cybern* 71 (1): 1–12.

Helmstaedter M, Briggman KL, Turaga SC, Jain V, Seung HS, et al. 2013. Connectomic reconstruction of the inner plexiform layer in the mouse retina. *Nature* 500 (7461): 168–174.

Hely MA, Reid WG, Adena MA, Halliday GM, and Morris JG. 2008. The Sydney multicenter study of Parkinson's disease: The inevitability of dementia at 20 years. *Movement Disorders* 23 (6): 837–844.

Hennig MH, Adams C, Willshaw D, and Sernagor E. 2009. Early-stage waves in the retinal network emerge close to a critical state transition between local and global functional connectivity. *J Neurosci* 29 (4): 1077–1086.

Henriksen S, Pang R, and Wronkiewicz M. 2016. A simple generative model of the mouse mesoscale connectome. *Elife* 5: e12366.

Hensch TK. 2004. Critical period regulation. *Annu Rev Neurosci* 27 (1): 549–579.

Hentschel HGE, and van Ooyen A. 1999. Models of axon guidance and bundling during development. *Proc R Soc Lond B Biol Sci* 266 (1434): 2231–2238.

Hentschel HGE, and van Ooyen A. 2000. Dynamic mechanisms for bundling and guidance during neural network formation. *Physica A: Statistical Mechanics and its Applications* 288 (1): 369–379.

Herculano-Houzel S. 2009. The human brain in numbers: A linearly scaled-up primate brain. *Front Hum Neurosci* 3: 31.

Herculano-Houzel S. 2017. Numbers of neurons as biological correlates of cognitive capability. *Curr Opin Behav Sci* 16: 1–7.

Herculano-Houzel S, Collins CE, Wong P, and Kaas JH. 2007. Cellular scaling rules for primate brains. *Proc Natl Acad Sci USA* 104 (9): 3562–3567.

Herculano-Houzel S, Collins CE, Wong P, Kaas JH, and Lent R. 2008. The basic nonuniformity of the cerebral cortex. *Proc Natl Acad Sci USA* 105 (34): 12593–12598.

Hermann B, Jones J, Sheth R, Dow C, Koehn M, et al. 2006. Children with new-onset epilepsy: Neuropsychological status and brain structure. *Brain* 129 (Pt 10): 2609–2619.

Heuer K, Gulban OF, Bazin PL, Osoianu A, Valabregue R, et al. 2019. Evolution of neocortical folding: A phylogenetic comparative analysis of MRI from 34 primate species. *Cortex* 118: 275–291.

Hilgetag CC, and Barbas H. 2006. Role of mechanical factors in the morphology of the primate cerebral cortex. *PLOS Comput Biol* 2 (3): e22.

Hilgetag CC, Beul SF, van Albada SJ, and Goulas A. 2019. An architectonic type principle integrates macroscopic cortico-cortical connections with intrinsic cortical circuits of the primate brain. *Netw Neurosci*: 1–40.

Hilgetag CC, Burns GA, O'Neill MA, Scannell JW, and Young MP. 2000. Anatomical connectivity defines the organization of clusters of cortical areas in the macaque monkey and the cat. *Philos Trans R Soc Lond B Biol Sci* 355 (1393): 91–110.

Hilgetag CC, and Kaiser M. 2004. Clustered organization of cortical connectivity. *Neuroinformatics* 2 (3): 353–360.

Hilgetag CC, O'Neill MA, and Young MP. 2000. Hierarchical organization of macaque and cat cortical sensory systems explored with a novel network processor. *Philos Trans R Soc Lond B Biol Sci* 355 (1393): 71–89.

Hill J, Inder T, Neil J, Dierker D, Harwell J, et al. 2010. Similar patterns of cortical expansion during human development and evolution. *Proc Natl Acad Sci USA* 107 (29): 13135–13140.

Hinton GE, and Sejnowski TJ. 1986. Learning and relearning in Boltzmann machines. In *Parallel distributed processing: Explorations in the microstructure of cognition* 1 (282–317): 2. Cambridge, MA: MIT Press.

His W. 1888. Zur geschichte des gehirns sowie der centralen und peripherischen nervenbahnen beim menschlichen embryo. *Abhandlungen der mathematisch-physikalischen Classe der Königlichl. Sächsichen Gesellschaft der Wissenschaften* 14 (7).

Holmes AJ, and Patrick LM. 2018. The myth of optimality in clinical neuroscience. *Trends Cogn Sci* 22 (3): 241–257.

Homae F, Watanabe H, Otobe T, Nakano T, Go T, et al. 2010. Development of global cortical networks in early infancy. *J Neurosci* 30 (14): 4877–4882.

Hong SJ, Bernhardt BC, Gill RS, Bernasconi N, and Bernasconi A. 2017. The spectrum of structural and functional network alterations in malformations of cortical development. *Brain* 140 (8): 2133–2143.

Hong SJ, Vos de Wael R, Bethlehem RAI, Lariviere S, Paquola C, et al. 2019. Atypical functional connectome hierarchy in autism. *Nat Commun* 10 (1): 1022.

Horn A, and Kuhn AA. 2015. Lead-DBS: A toolbox for deep brain stimulation electrode localizations and visualizations. *Neuroimage* 107: 127–135.

Hornykiewicz O, and Kish SJ. 1987. Biochemical pathophysiology of Parkinson's disease. *Adv Neurol* 45: 19–34.

Horton JC, and Adams DL. 2005. The cortical column: A structure without a function. *Philos Trans R Soc Lond B Biol Sci* 360 (1456): 837–862.

Hosoda C, Tanaka K, Nariai T, Honda M, and Hanakawa T. 2013. Dynamic neural network reorganization associated with second language vocabulary acquisition: A multimodal imaging study. *J Neurosci* 33 (34): 13663–13672.

Hubel DH, and Freeman DC. 1977. Projection into the visual field of ocular dominance columns in macaque monkey. *Brain Res* 122 (2): 336–343.

Hubel DH, Wiesel TN, and LeVay S. 1977. Plasticity of ocular dominance columns in monkey striate cortex. *Philos Trans R Soc Lond B Biol Sci* 278 (961): 377–409.

Hubel DH, Wiesel TN, and Stryker MP. 1977. Orientation columns in macaque monkey visual cortex demonstrated by the 2-deoxyglucose autoradiographic technique. *Nature* 269 (5626): 328–330.

Huberman AD. 2007. Mechanisms of eye-specific visual circuit development. *Curr Opin Neurobiol* 17 (1): 73–80.

Humphries MD, and Gurney K. 2008. Network 'small-world-ness': A quantitative method for determining canonical network equivalence. *PLOS One* 3 (4): e0002051.

Hutchings F, Thornton C, Zhang C, Wang Y, and Kaiser M. 2020. Predicting the impact of electric field stimulation in a detailed computational model of cortical tissue. *arXiv*:2001.10414.

Hutchinson E, Pulsipher D, Dabbs K, Myers y Gutierrez A, Sheth R, et al. 2010. Children with new-onset epilepsy exhibit diffusion abnormalities in cerebral white matter in the absence of volumetric differences. *Epilepsy Res* 88 (2–3): 208–214.

Huttenlocher PR. 1979. Synaptic density in human frontal cortex—developmental changes and effects of aging. *Brain Res* 163 (2): 195–205.

Huttenlocher PR. 1984. Synapse elimination and plasticity in developing human cerebral cortex. *Am J Ment Defic* 88 (5): 488–496.

Ingalhalikar M, Smith A, Parker D, Satterthwaite TD, Elliott MA, et al. 2014. Sex differences in the structural connectome of the human brain. *Proc Natl Acad Sci USA* 111 (2): 823–828.

Innocenti GM, and Price DJ. 2005. Exuberance in the development of cortical networks. *Nat Rev Neurosci* 6 (12): 955–965.

Innocenti GM, Vercelli A, and Caminiti R. 2014. The diameter of cortical axons depends both on the area of origin and target. *Cereb Cortex* 24 (8): 2178–2188.

Isingrini M, and Vazou F. 1997. Relation between fluid intelligence and frontal lobe functioning in older adults. *Int J Aging Hum Dev* 45 (2): 99–109.

Ito M, Masuda N, Shinomiya K, Endo K, and Ito K. 2013. Systematic analysis of neural projections reveals clonal composition of the *Drosophila* brain. *Curr Biol* 23 (8): 644–655.

Iturria-Medina Y, Sotero RC, Canales-Rodriguez EJ, Aleman-Gomez Y, and Melie-Garcia L. 2008. Studying the human brain anatomical network via diffusion-weighted MRI and graph theory. *Neuroimage* 40 (3): 1064–1076.

Jackson GM, Draper A, Dyke K, Pepes SE, and Jackson SR. 2015. Inhibition, disinhibition, and the control of action in Tourette syndrome. *Trends Cogn Sci* 19 (11): 655–665.

Jackson SR, Parkinson A, Jung J, Ryan SE, Morgan PS, et al. 2011. Compensatory neural reorganization in Tourette syndrome. *Curr Biol* 21 (7): 580–585.

References

Jacobs RA, and Jordan MI. 1992. Computational consequences of a bias towards short connections. *J Cogn Neurosci* 4: 323–336.

Jardim-Messeder D, Lambert K, Noctor S, Pestana FM, de Castro Leal ME, et al. 2017. Dogs have the most neurons, though not the largest brain: Trade-off between body mass and number of neurons in the cerebral cortex of large carnivoran species. *Front Neuroanat* 11: 118.

Jbabdi S, and Johansen-Berg H. 2011. Tractography: Where do we go from here? *Brain Conn* 1 (3): 169–183.

Jbabdi S, Sotiropoulos SN, Haber SN, Van Essen DC, and Behrens TE. 2015. Measuring macroscopic brain connections *in vivo*. *Nat Neurosci* 18 (11): 1546–1555.

Jiang L, Xu Y, Zhu XT, Yang Z, Li HJ, et al. 2015. Local-to-remote cortical connectivity in early- and adulthood-onset schizophrenia. *Transl Psychiatry* 5: e566.

Jiang X, Shen Y, Yao J, Zhang L, Xu L, et al. 2019. Connectome analysis of functional and structural hemispheric brain networks in major depressive disorder. *Transl Psychiatry* 9 (1): 136.

Johansen-Berg H. 2007. Structural plasticity: Rewiring the brain. *Curr Biol* 17 (4): R141–144.

Johansen-Berg H, Scholz J, and Stagg CJ. 2010. Relevance of structural brain connectivity to learning and recovery from stroke. *Front Syst Neurosci* 4: 146.

Jones DK. 2010. Challenges and limitations of quantifying brain connectivity in vivo with diffusion MRI. *Imaging* 2 (3): 341–355.

Jones DT, Knopman DS, Gunter JL, Graff-Radford J, Vemuri P, et al. 2016. Cascading network failure across the Alzheimer's disease spectrum. *Brain* 139 (Pt 2): 547–562.

Jucker M, and Walker LC. 2013. Self-propagation of pathogenic protein aggregates in neurodegenerative diseases. *Nature* 501 (7465): 45–51.

Jun SB. 2012. Ultrasound as a noninvasive neuromodulation tool. *Biomed Eng Lett* 2 (1): 8–12.

Jung RE, and Haier RJ. 2007. The parieto-frontal integration theory (P-FIT) of intelligence: Converging neuroimaging evidence. *Behav Brain Sci* 30 (2): 135–154.

Kaiser M. 2007. Brain architecture: A design for natural computation. *Philos Trans A Math Phys Eng Sci* 365 (1861): 3033–3045.

Kaiser M. 2008. Mean clustering coefficients: The role of isolated nodes and leafs on clustering measures for small-world networks. *New J Phys* 10: 083042.

Kaiser M. 2011. A tutorial in connectome analysis: Topological and spatial features of brain networks. *Neuroimage* 57 (3): 892–907.

Kaiser M. 2013. The potential of the human connectome as a biomarker of brain disease. *Front Hum Neurosci* 7: 484.

Kaiser M. 2015. Neuroanatomy: Connectome connects fly and mammalian brain networks. *Curr Biol* 25 (10): R416–418.

Kaiser M. 2017. Mechanisms of connectome development. *Trends Cogn Sci* 21 (9): 703–717.

Kaiser M, Görner M, and Hilgetag CC. 2007. Functional criticality in clustered networks without inhibition. *New J Phys* 9: 110.

Kaiser M, and Hilgetag CC. 2004a. Modelling the development of cortical systems networks. *Neurocomputing* 58–60: 297–302.

Kaiser M, and Hilgetag CC. 2004b. Spatial growth of real-world networks. *Phys Rev E Stat Nonlin Soft Matter Phys* 69 (3 Pt 2): 036103.

Kaiser M, and Hilgetag CC. 2006. Nonoptimal component placement, but short processing paths, due to long-distance projections in neural systems. *PLOS Comput Biol* 2 (7): e95.

Kaiser M, and Hilgetag CC. 2007. Development of multi-cluster cortical networks by time windows for spatial growth. *Neurocomputing* 70 (10–12): 1829–1832.

Kaiser M, and Hilgetag CC. 2010. Optimal hierarchical modular topologies for producing limited sustained activation of neural networks. *Front Neuroinform* 4: 8.

Kaiser M, Hilgetag CC, and Kötter R. 2010. Hierarchy and dynamics of neural networks. *Front Neuroinform* 4: 112.

Kaiser M, Hilgetag CC, and van Ooyen A. 2009. A simple rule for axon outgrowth and synaptic competition generates realistic connection lengths and filling fractions. *Cereb Cortex* 19 (12): 3001–3010.

Kaiser M, Martin R, Andras P, and Young MP. 2007. Simulation of robustness against lesions of cortical networks. *Eur J Neurosci* 25 (10): 3185–3192.

Kalisman N, Silberberg G, and Markram H. 2005. The neocortical microcircuit as a tabula rasa. *Proc Natl Acad Sci USA* 102 (3): 880–885.

Karbowski J. 2001. Optimal wiring principle and plateaus in the degree of separation for cortical neurons. *Phys Rev Lett* 86 (16): 3674–3677.

Katz LC, and Crowley JC. 2002. Development of cortical circuits: Lessons from ocular dominance columns. *Nat Rev Neurosci* 3 (1): 34–42.

Katzman R. 1993. Education and the prevalence of dementia and Alzheimer's disease. *Neurology* 43: 13–20.

Katzman R, Terry R, DeTeresa R, Brown T, Davies P, et al. 1988. Clinical, pathological, and neurochemical changes in dementia: A subgroup with preserved mental status and numerous neocortical plaques. *Annals of Neurology* 23 (2): 138–144.

Kaufman A, Dror G, Meilijson I, and Ruppin E. 2006. Gene expression of *C. elegans* neurons carries information on their synaptic connectivity. *PLOS Computational Biology* 2 (12): e167.

Kaufmann T, Alnaes D, Doan NT, Brandt CL, Andreassen OA, et al. 2017. Delayed stabilization and individualization in connectome development are related to psychiatric disorders. *Nat Neurosci* 20 (4): 513–515.

Keck T, Mrsic-Flogel TD, Vaz Afonso M, Eysel UT, Bonhoeffer T, et al. 2008. Massive restructuring of neuronal circuits during functional reorganization of adult visual cortex. *Nat Neurosci* 11 (10): 1162–1167.

Kekic M, Boysen E, Campbell IC, and Schmidt U. 2016. A systematic review of the clinical efficacy of transcranial direct current stimulation (tDCS) in psychiatric disorders. *J Psychiatr Res* 74: 70–86.

Kelava I, Rentzsch F, and Technau U. 2015. Evolution of eumetazoan nervous systems: Insights from cnidarians. *Philos Trans R Soc Lond B Biol Sci* 370 (1684).

Keller CJ, Huang Y, Herrero JL, Fini ME, Du V, et al. 2018. Induction and quantification of excitability changes in human cortical networks. *J Neurosci* 38 (23): 5384–5398.

Kelsch W, Sim S, and Lois C. 2010. Watching synaptogenesis in the adult brain. *Annu Rev Neurosci* 33: 131–149.

Kendler KS, Aggen SH, Knudsen GP, Roysamb E, Neale MC, et al. 2011. The structure of genetic and environmental risk factors for syndromal and subsyndromal common DSM-IV axis I and all axis II disorders. *Am J Psychiatry* 168 (1): 29–39.

Khambhati AN, Kahn AE, Costantini J, Ezzyat Y, Solomon EA, et al. 2019. Functional control of electrophysiological network architecture using direct neurostimulation in humans. *Netw Neurosci*: 1–30.

Kim HR, Lee P, Seo SW, Roh JH, Oh M, et al. 2019. Comparison of amyloid beta and tau spread models in Alzheimer's disease. *Cereb Cortex* 29 (10): 4291–4302.

Kim JS, and Kaiser M. 2014. From *Caenorhabditis elegans* to the human connectome: A specific modular organization increases metabolic, functional and developmental efficiency. *Philos Trans R Soc Lond B Biol Sci* 369 (1653): 20130529.

Klein D, Rotarska-Jagiela A, Genc E, Sritharan S, Mohr H, et al. 2014. Adolescent brain maturation and cortical folding: Evidence for reductions in gyrification. *PLOS One* 9 (1): e84914.

Koch C. 2004. *Biophysics of computation: Information processing in single neurons.* New York: Oxford University Press.

Koene RA, Tijms B, van Hees P, Postma F, de Ridder A, et al. 2009. Netmorph: A framework for the stochastic generation of large scale neuronal networks with realistic neuron morphologies. *Neuroinformatics* 7 (3): 195–210.

Kohonen TK. 1982. Self-organized formation of topologically correct feature maps. *Biol Cybern* 43: 41–65.

König P, Engel AK, Roelfsema PR, and Singer W. 1995. How precise is neuronal synchronization? *Neural Comput* 7 (3): 469–485.

Koser DE, Thompson AJ, Foster SK, Dwivedy A, Pillai EK, et al. 2016. Mechanosensing is critical for axon growth in the developing brain. *Nat Neurosci* 19 (12): 1592–1598.

Kötter R. 2004. Online retrieval, processing, and visualization of primate connectivity data from the CoCoMac database. *Neuroinformatics* 2 (2): 127–144.

Krubitzer L, and Kahn DM. 2003. Nature versus nurture revisited: An old idea with a new twist. *Prog Neurobiol* 70 (1): 33–52.

Kuan C-Y, Roth KA, Flavell RA, and Rakic P. 2000. Mechanisms of programmed cell death in the developing brain. *Trends Neurosci* 23 (7): 291–297.

Kubanek J, Shukla P, Das A, Baccus SA, and Goodman MB. 2018. Ultrasound elicits behavioral responses through mechanical effects on neurons and ion channels in a simple nervous system. *J Neurosci* 38 (12): 3081–3091.

LaMantia AS, and Rakic P. 1990. Axon overproduction and elimination in the corpus callosum of the developing rhesus monkey. *J Neurosci* 10 (7): 2156–2175.

LaMantia AS, and Rakic P. 1994. Axon overproduction and elimination in the anterior commissure of the developing rhesus monkey. *J Comp Neurol* 340 (3): 328–336.

Larson TA, Thatra NM, Lee BH, and Brenowitz EA. 2014. Reactive neurogenesis in response to naturally occurring apoptosis in an adult brain. *J Neurosci* 34 (39): 13066–13076.

Latora V, and Marchiori M. 2001. Efficient behavior of small-world networks. *Phys Rev Lett* 87 (19): 198701.

Lau YC, Hinkley LB, Bukshpun P, Strominger ZA, Wakahiro ML, et al. 2013. Autism traits in individuals with agenesis of the corpus callosum. *J Autism Dev Disord* 43 (5): 1106–1118.

Laughlin SB, and Sejnowski TJ. 2003. Communication in neuronal networks. *Science* 301 (5641): 1870–1874.

Lebel C, and Beaulieu C. 2011. Longitudinal development of human brain wiring continues from childhood into adulthood. *J Neurosci* 31 (30): 10937–10947.

Legon W, Ai L, Bansal P, and Mueller JK. 2018. Neuromodulation with single-element transcranial focused ultrasound in human thalamus. *Hum Brain Mapp* 39 (5): 1995–2006.

Leinenga G, Langton C, Nisbet R, and Gotz J. 2016. Ultrasound treatment of neurological diseases—current and emerging applications. *Nat Rev Neurol* 12 (3): 161–174.

Lemaire T, Neufeld E, Kuster N, and Micera S. 2019. Understanding ultrasound neuromodulation using a computationally efficient and interpretable model of intramembrane cavitation. *J Neural Eng* 16 (4): 046007.

Lemieux L, Daunizeau J, and Walker MC. 2011. Concepts of connectivity and human epileptic activity. *Front Syst Neurosci* 5: 12.

Lewis JD, Evans AC, Pruett JR, Jr., Botteron KN, McKinstry RC, et al. 2017. The emergence of network inefficiencies in infants with autism spectrum disorder. *Biol Psychiatry* 82 (3): 176–185.

Li A, Gong H, Zhang B, Wang Q, Yan C, et al. 2010. Micro-optical sectioning tomography to obtain a high-resolution atlas of the mouse brain. *Science* 330 (6009): 1404–1408.

Li HJ, Xu Y, Zhang KR, Hoptman MJ, and Zuo XN. 2015. Homotopic connectivity in drug-naive, first-episode, early-onset schizophrenia. *J Child Psychol Psychiatry* 56 (4): 432–443.

Li LM, Violante IR, Leech R, Hampshire A, Opitz A, et al. 2019. Cognitive enhancement with salience network electrical stimulation is influenced by network structural connectivity. *Neuroimage* 185: 425–433.

Li WC, Cooke T, Sautois B, Soffe SR, Borisyuk R, et al. 2007. Axon and dendrite geography predict the specificity of synaptic connections in a functioning spinal cord network. *Neural Dev* 2 (1): 17.

Lichtman JW, Livet J, and Sanes JR. 2008. A technicolour approach to the connectome. *Nat Rev Neurosci* 9 (6): 417–422.

Lim S, Han CE, Uhlhaas PJ, and Kaiser M. 2015. Preferential detachment during human brain development: Age- and sex-specific structural connectivity in diffusion tensor imaging (DTI) data. *Cereb Cortex* 25 (6): 1477–1489.

Lim S, and Kaiser M. 2015. Developmental time windows for axon growth influence neuronal network topology. *Biol Cybern* 109 (2): 275–286.

Lin HY, Perry A, Cocchi L, Roberts JA, Tseng WI, et al. 2019. Development of frontoparietal connectivity predicts longitudinal symptom changes in young people with autism spectrum disorder. *Transl Psychiatry* 9 (1): 86.

Lin MK, Takahashi YS, Huo BX, Hanada M, Nagashima J, et al. 2019. A high-throughput neurohistological pipeline for brain-wide mesoscale connectivity mapping of the common marmoset. *Elife* 8.

Little GE, Lopez-Bendito G, Runker AE, Garcia N, Pinon MC, et al. 2009. Specificity and plasticity of thalamocortical connections in *Sema6A* mutant mice. *PLOS Biol* 7 (4): e98.

Liu Y, Liang M, Zhou Y, He Y, Hao Y, et al. 2008. Disrupted small-world networks in schizophrenia. *Brain* 131 (Pt 4): 945–961.

Liu YY, Slotine JJ, and Barabasi AL. 2011. Controllability of complex networks. *Nature* 473 (7346): 167–173.

Lo C-Y, Wang P-N, Chou K-H, Wang J, He Y, et al. 2010. Diffusion tensor tractography reveals abnormal topological organization in structural cortical networks in Alzheimer's disease. *J Neurosci* 30 (50): 16876–16885.

Lohof AM, Delhaye-Bouchaud N, and Mariani J. 1996. Synapse elimination in the central nervous system: Functional significance and cellular mechanisms. *Rev Neurosci* 7 (2): 85–101.

Long P, and Corfas G. 2014. Dynamic regulation of myelination in health and disease. *JAMA Psychiatry* 71 (11): 1296–1297.

Lorenz KZ. 1958. The evolution of behavior. *Sci Am* (199): 67–74.

Lövdén M, Xu W, and Wang H-X. 2013. Lifestyle change and the prevention of cognitive decline and dementia: What is the evidence? *Curr Opin Psychiatry* 26 (3): 239–243.

Luders E, Kurth F, Mayer EA, Toga AW, Narr KL, et al. 2012. The unique brain anatomy of meditation practitioners: Alterations in cortical gyrification. *Front Hum Neurosci* 6: 34.

Luders E, Narr KL, Thompson PM, Rex DE, Jancke L, et al. 2004. Gender differences in cortical complexity. *Nat Neurosci* 7 (8): 799–800.

Luders E, Thompson PM, Narr KL, Toga AW, Jancke L, et al. 2006. A curvature-based approach to estimate local gyrification on the cortical surface. *Neuroimage* 29 (4): 1224–1230.

Luft CD, Pereda E, Banissy MJ, and Bhattacharya J. 2014. Best of both worlds: Promise of combining brain stimulation and brain connectome. *Front Syst Neurosci* 8: 132.

Lund JS, Angelucci A, and Bressloff PC. 2003. Anatomical substrates for functional columns in macaque monkey primary visual cortex. *Cereb Cortex* 13 (1): 15–24.

Luo L, and O'Leary DD. 2005. Axon retraction and degeneration in development and disease. *Annu Rev Neurosci* 28: 127–156.

Madan CR, and Kensinger EA. 2016. Cortical complexity as a measure of age-related brain atrophy. *Neuroimage* 134: 617–629.

Majka P, Chaplin TA, Yu HH, Tolpygo A, Mitra PP, et al. 2016. Towards a comprehensive atlas of cortical connections in a primate brain: Mapping tracer injection studies of the common marmoset into a reference digital template. *J Comp Neurol* 524 (11): 2161–2181.

Mallamaci A, and Stoykova A. 2006. Gene networks controlling early cerebral cortex arealization. *Eur J Neurosci* 23 (4): 847–856.

Mander BA, Marks SM, Vogel JW, Rao V, Lu B, et al. 2015. Beta-amyloid disrupts human NREM slow waves and related hippocampus-dependent memory consolidation. *Nat Neurosci* 18 (7): 1051–1057.

Manger PR, Restrepo CE, and Innocenti GM. 2010. The superior colliculus of the ferret: Cortical afferents and efferent connections to dorsal thalamus. *Brain Res* 1353: 74–85.

Markov NT, Ercsey-Ravasz M, Lamy C, Ribeiro Gomes AR, Magrou L, et al. 2013. The role of long-range connections on the specificity of the macaque interareal cortical network. *Proc Natl Acad Sci USA* 110 (13): 5187–5192.

Markov NT, Ercsey-Ravasz M, Van Essen DC, Knoblauch K, Toroczkai Z, et al. 2013. Cortical high-density counterstream architectures. *Science* 342 (6158): 1238406.

References

Markov NT, Ercsey-Ravasz MM, Ribeiro Gomes AR, Lamy C, Magrou L, et al. 2014. A weighted and directed interareal connectivity matrix for macaque cerebral cortex. *Cereb Cortex* 24 (1): 17–36.

Masuda N, and Aihara K. 2004. Global and local synchrony of coupled neurons in small-world networks. *Biol Cybern* 90 (4): 302–309.

McKeith I, O'Brien J, Walker Z, Tatsch K, Booij J, et al. 2007. Sensitivity and specificity of dopamine transporter imaging with 123I-FP-CIT SPECT in dementia with Lewy bodies: A phase III, multicentre study. *Lancet Neurol* 6 (4): 305–313.

McKeith IG, Boeve BF, Dickson DW, Halliday G, Taylor JP, et al. 2017. Diagnosis and management of dementia with Lewy bodies: Fourth consensus report of the DLB consortium. *Neurology* 89 (1): 88–100.

McKeith IG, Dickson DW, Lowe J, Emre M, O'Brien JT, et al. 2005. Diagnosis and management of dementia with Lewy bodies: Third report of the DLB consortium. *Neurology* 65 (12): 1863–1872.

Meinhardt H. 2008. Models of biological pattern formation: From elementary steps to the organization of embryonic axes. *Curr Top Dev Biol* 81: 1–63.

Metropolis N, Rosenbluth AW, Rosenbluth MN, Teller AH, and Teller E. 1953. Equation of state calculations by fast computing machines. *J Chem Phys* 21: 1087–1092.

Meunier D, Lambiotte R, and Bullmore ET. 2010. Modular and hierarchically modular organization of brain networks. *Front Neurosci* 4: 200.

Meynert T. 1867. Der bau der grosshirnrinde und seiner örtlichen verschiedenheiten, nebst einem pathologisch-anatomischen collarium. *Vierteljahresschr Psychiat* 1: 77–93.

Mezias C, Rey N, Brundin P, and Raj A. 2020. Neural connectivity predicts spreading of alpha-synuclein pathology in fibril-injected mouse models: Involvement of retrograde and anterograde axonal propagation. *Neurobiology of Disease*: 104623.

Micheloyannis S, Pachou E, Stam CJ, Breakspear M, Bitsios P, et al. 2006. Small-world networks and disturbed functional connectivity in schizophrenia. *Schizophr Res* 87 (1–3): 60–66.

Mikula S, Stone JM, and Jones EG. 2008. Brainmaps.Org—interactive high-resolution digital brain atlases and virtual microscopy. *Brains Minds Media* 3: bmm1426.

Milgram S. 1967. The small-world problem. *Psychology Today* 1: 60–67.

Milham MP, Ai L, Koo B, Xu T, Amiez C, et al. 2018. An open resource for non-human primate imaging. *Neuron* 100 (1): 61–74 e62.

Miller KL, Alfaro-Almagro F, Bangerter NK, Thomas DL, Yacoub E, et al. 2016. Multimodal population brain imaging in the UK Biobank prospective epidemiological study. *Nat Neurosci* 19 (11): 1523–1536.

Milo R, Shen-Orr S, Itzkovitz S, Kashtan N, Chklovskii D, et al. 2002. Network motifs: Simple building blocks of complex networks. *Science* 298 (5594): 824–827.

Minerbi A, Kahana R, Goldfeld L, Kaufman M, Marom S, et al. 2009. Long-term relationships between synaptic tenacity, synaptic remodeling, and network activity. *PLOS Biol* 7 (6): e1000136.

Mohades SG, Struys E, Van Schuerbeek P, Mondt K, Van De Craen P, et al. 2012. DTI reveals structural differences in white matter tracts between bilingual and monolingual children. *Brain Res* 1435: 72–80.

Mohan UR, Watrous AJ, Miller JF, Lega BC, Sperling MR, et al. 2019. The effects of direct brain stimulation in humans depend on frequency, amplitude, and white-matter proximity. *bioRxiv* 746834, doi:10.1101/746834.

Monod J. 1949. The growth of bacterial cultures. *Annu Rev Microbiol* 3: 371–394.

Mota B, and Herculano-Houzel S. 2012. How the cortex gets its folds: An inside-out, connectivity-driven model for the scaling of mammalian cortical folding. *Front Neuroanat* 6: 3.

Mota B, and Herculano-Houzel S. 2015. Cortical folding scales universally with surface area and thickness, not number of neurons. *Science* 349 (6243): 74.

Mountcastle VB. 1997. The columnar organization of the neocortex. *Brain* 120 (Pt 4): 701–722.

Muir DR, Da Costa NM, Girardin CC, Naaman S, Omer DB, et al. 2011. Embedding of cortical representations by the superficial patch system. *Cereb Cortex* 21 (10): 2244–2260.

Murray JD. 2003. *Mathematical biology. II: Spatial models and biomedical applications*. 3rd ed. Heidelberg: Springer.

Murre JM, and Sturdy DP. 1995. The connectivity of the brain: Multi-level quantitative analysis. *Biol Cybern* 73 (6): 529–545.

Natu VS, Gomez J, Barnett M, Jeska B, Kirilina E, et al. 2019. Apparent thinning of human visual cortex during childhood is associated with myelination. *Proc Natl Acad Sci USA* 116 (41): 20750–20759.

Nawa H, Takahashi M, and Patterson PH. 2000. Cytokine and growth factor involvement in schizophrenia—support for the developmental model. *Molecular Psychiatry* 5 (6): 594–603.

Nepusz T, Négyessy L, Tusnády G, and Bazsó F. 2008. Reconstructing cortical networks: Case of directed graphs with high level of reciprocity. In *Handbook of Large-Scale Random Networks*, 325–368. Berlin: Springer.

Neumann WJ, Schroll H, de Almeida Marcelino AL, Horn A, Ewert S, et al. 2018. Functional segregation of basal ganglia pathways in Parkinson's disease. *Brain* 141 (9): 2655–2669.

Newman ME. 2004. Fast algorithm for detecting community structure in networks. *Phys Rev E Stat Nonlin Soft Matter Phys* 69 (6 Pt 2): 066133.

References

Newman ME. 2006. Modularity and community structure in networks. *Proc Natl Acad Sci USA* 103 (23): 8577–8582.

Ng YS, van Ruiten H, Lai HM, Scott R, Ramesh V, et al. 2018. The adjunctive application of transcranial direct current stimulation in the management of de novo refractory epilepsia partialis continua in adolescent-onset *POLG*-related mitochondrial disease. *Epilepsia Open* 3 (1): 103–108.

Nie J, Guo L, Li G, Faraco C, Stephen Miller L, et al. 2010. A computational model of cerebral cortex folding. *J Theor Biol* 264 (2): 467–478.

Nie J, Guo L, Li K, Wang Y, Chen G, et al. 2012. Axonal fiber terminations concentrate on gyri. *Cereb Cortex* 22 (12): 2831–2839.

Nikolic K, Jarvis S, Grossman N, and Schultz S. 2013. "Computational models of optogenetic tools for controlling neural circuits with light." Annual International Conference of the IEEE Engineering in Medicine and Biology Society.

Nisbach F, and Kaiser M. 2007. Developmental time windows for spatial growth generate multiple-cluster small-world networks. *Eur Phys J B* 58 (2): 185–191.

Nitsche MA, Cohen LG, Wassermann EM, Priori A, Lang N, et al. 2008. Transcranial direct current stimulation: State of the art 2008. *Brain Stimul* 1 (3): 206–223.

Nitsche MA, Muller-Dahlhaus F, Paulus W, and Ziemann U. 2012. The pharmacology of neuroplasticity induced by non-invasive brain stimulation: Building models for the clinical use of CNS active drugs. *J Physiol* 590 (19): 4641–4662.

Nitsche MA, Schauenburg A, Lang N, Liebetanz D, Exner C, et al. 2003. Facilitation of implicit motor learning by weak transcranial direct current stimulation of the primary motor cortex in the human. *J Cogn Neurosci* 15 (4): 619–626.

Nordahl CW, Dierker D, Mostafavi I, Schumann CM, Rivera SM, et al. 2007. Cortical folding abnormalities in autism revealed by surface-based morphometry. *J Neurosci* 27 (43): 11725–11735.

Nottebohm F. 2005. The neural basis of birdsong. *PLOS Biol* 3 (5): e164.

Oh SW, Harris JA, Ng L, Winslow B, Cain N, et al. 2014. A mesoscale connectome of the mouse brain. *Nature* 508 (7495): 207–214.

Okujeni S, and Egert U. 2019. Self-organization of modular network architecture by activity-dependent neuronal migration and outgrowth. *eLife* 8: e47996.

Palla G, Derenyi I, Farkas I, and Vicsek T. 2005. Uncovering the overlapping community structure of complex networks in nature and society. *Nature* 435 (7043): 814–818.

Palmqvist S, Hansson O, Minthon L, and Londos E. 2009. Practical suggestions on how to differentiate dementia with Lewy bodies from Alzheimer's disease with common cognitive tests. *Int J Geriatr Psychiatry* 24 (12): 1405–1412.

Pandya S, Zeighami Y, Freeze B, Dadar M, Collins DL, et al. 2019. Predictive model of spread of Parkinson's pathology using network diffusion. *Neuroimage* 192: 178–194.

Pang T, Atefy R, and Sheen V. 2008. Malformations of cortical development. *Neurologist* 14 (3): 181–191.

Passingham R. 2009. How good is the macaque monkey model of the human brain? *Curr Opin Neurobiol* 19 (1): 6–11.

Paul LK. 2011. Developmental malformation of the corpus callosum: A review of typical callosal development and examples of developmental disorders with callosal involvement. *J Neurodev Disord* 3 (1): 3–27.

Payne BR, and Lomber SG. 2001. Reconstructing functional systems after lesions of cerebral cortex. *Nat Rev Neurosci* 2 (12): 911–919.

Peraza LR, Díaz-Parra A, Kennion O, Moratal D, Taylor J-P, et al. 2019. Structural connectivity centrality changes mark the path toward Alzheimer's disease. *Alzheimer's & Dementia: Diagnosis, Assessment & Disease Monitoring* 11: 98–107.

Pereira JB, Mijalkov M, Kakaei E, Mecocci P, Vellas B, et al. 2016. Disrupted network topology in patients with stable and progressive mild cognitive impairment and Alzheimer's disease. *Cereb Cortex* 26 (8): 3476–3493.

Perin R, Berger TK, and Markram H. 2011. A synaptic organizing principle for cortical neuronal groups. *Proc Natl Acad Sci USA* 108 (13): 5419–5424.

Petanjek Z, Judas M, Simic G, Rasin MR, Uylings HB, et al. 2011. Extraordinary neoteny of synaptic spines in the human prefrontal cortex. *Proc Natl Acad Sci USA* 108 (32): 13281–13286.

Pettersson-Yeo W, Allen P, Benetti S, McGuire P, and Mechelli A. 2011. Dysconnectivity in schizophrenia: Where are we now? *Neuroscience & Biobehavioral Reviews* 35 (5): 1110–1124.

Picco N, Garcia-Moreno F, Maini PK, Woolley TE, and Molnar Z. 2018. Mathematical modeling of cortical neurogenesis reveals that the founder population does not necessarily scale with neurogenic output. *Cereb Cortex* 28 (7): 2540–2550.

Picker A, Cavodeassi F, Machate A, Bernauer S, Hans S, et al. 2009. Dynamic coupling of pattern formation and morphogenesis in the developing vertebrate retina. *PLOS Biol* 7 (10): e1000214.

Polania R, Nitsche MA, and Paulus W. 2011. Modulating functional connectivity patterns and topological functional organization of the human brain with transcranial direct current stimulation. *Hum Brain Mapp* 32 (8): 1236–1249.

Price CJ, and Friston KJ. 2002. Degeneracy and cognitive anatomy. *Trends Cogn Sci* 6 (10): 416–421.

Purves D, and Lichtman JW. 1980. Elimination of synapses in the developing nervous system. *Science* 210 (4466): 153–157.

Putcha GV, and Johnson EM, Jr. 2004. Men are but worms: Neuronal cell death in *C. elegans* and vertebrates. *Cell Death Differ* 11 (1): 38–48.

Qubbaj MR, and Jirsa VK. 2007. Neural field dynamics with heterogeneous connection topology. *Phys Rev Lett* 98 (23): 238102.

Raj A, Kuceyeski A, and Weiner M. 2012. A network diffusion model of disease progression in dementia. *Neuron* 73 (6): 1204–1215.

Raj A, LoCastro E, Kuceyeski A, Tosun D, Relkin N, et al. 2015. Network diffusion model of progression predicts longitudinal patterns of atrophy and metabolism in Alzheimer's disease. *Cell Rep* 10 (3): 359–369.

Rakic P. 1986. Mechanism of ocular dominance segregation in the lateral geniculate nucleus— competitive elimination hypothesis. *Trends Neurosci* 9 (1): 11–15.

Rakic P. 1995. A small step for the cell, a giant leap for mankind: A hypothesis of neocortical expansion during evolution. *Trends Neurosci* 18 (9): 383–388.

Rakic P. 2002. Neurogenesis in adult primate neocortex: An evaluation of the evidence. *Nat Rev Neurosci* 3 (1): 65–71.

Rakic P. 2008. Confusing cortical columns. *Proc Natl Acad Sci USA* 105 (34): 12099–12100.

Rakic P, Bourgeois JP, Eckenhoff MF, Zecevic N, and Goldman-Rakic PS. 1986. Concurrent overproduction of synapses in diverse regions of the primate cerebral cortex. *Science* 232 (4747): 232–235.

Rakic P, and Riley KP. 1983. Overproduction and elimination of retinal axons in the fetal rhesus monkey. *Science* 219 (4591): 1441–1444.

Rakic S, and Zecevic N. 2000. Programmed cell death in the developing human telencephalon. *Eur J Neurosci* 12 (8): 2721–2734.

Ramón y Cajal S. 1892. *The structure of the retina*. Translated by Thorpe, SA and Glickstein, M., 1972. Springfield, IL: Thomas.

Ravasz E, Somera AL, Mongru DA, Oltvai ZN, and Barabasi AL. 2002. Hierarchical organization of modularity in metabolic networks. *Science* 297 (5586): 1551–1555.

Reato D, Bikson M, and Parra LC. 2015. Lasting modulation of in vitro oscillatory activity with weak direct current stimulation. *J Neurophysiol* 113 (5): 1334–1341.

Reichert H. 1990. *Neurobiologie*. Stuttgart: Thieme.

Reijneveld JC, Ponten SC, Berendse HW, and Stam CJ. 2007. The application of graph theoretical analysis to complex networks in the brain. *Clin Neurophysiol* 118 (11): 2317–2331.

Reinhart RMG, and Nguyen JA. 2019. Working memory revived in older adults by synchronizing rhythmic brain circuits. *Nat Neurosci* 22 (5): 820–827.

Ribeiro P, Silva F, and Kaiser M. 2009. "Strategies for network motifs discovery." IEEE Conference on e-Science, Oxford.

Riddle D, Blumenthal T, Meyer B, and Priess J, eds. 1997. *C. elegans II*. Cold Spring Harbor, NY: Cold Spring Harbor Press.

Riley EP, Infante MA, and Warren KR. 2011. Fetal alcohol spectrum disorders: An overview. *Neuropsychol Rev* 21 (2): 73–80.

Robinson PA. 2013. Neural field theory with variance dynamics. *J Math Biol* 66 (7): 1475–1497.

Rockel AJ, Hiorns RW, and Powell TP. 1980. The basic uniformity in structure of the neocortex. *Brain* 103 (2): 221–244.

Rockland KS, and Lund JS. 1982. Widespread periodic intrinsic connections in the tree shrew visual cortex. *Science* 215 (4539): 1532.

Ronan L, and Fletcher PC. 2015. From genes to folds: A review of cortical gyrification theory. *Brain Struct Funct* 220 (5): 2475–2483.

Ropireddy D, and Ascoli GA. 2011. Potential synaptic connectivity of different neurons onto pyramidal cells in a 3D reconstruction of the rat hippocampus. *Front Neuroinform* 5: 5.

Rubenstein JL, and Merzenich MM. 2003. Model of autism: Increased ratio of excitation/inhibition in key neural systems. *Genes Brain Behav* 2 (5): 255–267.

Rubinov M, and Bullmore E. 2013. Schizophrenia and abnormal brain network hubs. *Dialogues Clin Neurosci* 15 (3): 339–349.

Rubinov M, McIntosh AR, Valenzuela MJ, and Breakspear M. 2009. Simulation of neuronal death and network recovery in a computational model of distributed cortical activity. *Am J Geriatr Psychiatry* 17 (3): 210–217.

Rubinov M, and Sporns O. 2010. Complex network measures of brain connectivity: Uses and interpretations. *Neuroimage* 52 (3): 1059–1069.

Ryan K, Lu Z, and Meinertzhagen IA. 2016. The CNS connectome of a tadpole larva of *Ciona intestinalis* (L.) highlights siddedness in the brain of a chordate sibling. *Elife* 5.

Sabuncu MR, Ge T, Holmes AJ, Smoller JW, Buckner RL, et al. 2016. Morphometricity as a measure of the neuroanatomical signature of a trait. *Proc Natl Acad Sci USA* 113 (39): E5749–E5756.

Sadek AR, Magill PJ, and Bolam JP. 2007. A single-cell analysis of intrinsic connectivity in the rat globus pallidus. *J Neurosci* 27 (24): 6352–6362.

Salami M, Itami C, Tsumoto T, and Kimura F. 2003. Change of conduction velocity by regional myelination yields constant latency irrespective of distance between thalamus and cortex. *Proc Natl Acad Sci USA* 100 (10): 6174–6179.

Sallet PC, Elkis H, Alves TM, Oliveira JR, Sassi E, et al. 2003. Reduced cortical folding in schizophrenia: An MRI morphometric study. *Am J Psychiatry* 160 (9): 1606–1613.

References

Sanchez-Rodriguez LM, Iturria-Medina Y, Baines EA, Mallo SC, Dousty M, et al. 2018. Design of optimal nonlinear network controllers for Alzheimer's disease. *PLOS Comput Biol* 14 (5): e1006136.

Scannell JW, Blakemore C, and Young MP. 1995. Analysis of connectivity in the cat cerebral cortex. *J Neurosci* 15 (2): 1463–1483.

Scannell JW, Burns GA, Hilgetag CC, O'Neil MA, and Young MP. 1999. The connectional organization of the cortico-thalamic system of the cat. *Cereb Cortex* 9 (3): 277–299.

Schlegel AA, Rudelson JJ, and Tse PU. 2012. White matter structure changes as adults learn a second language. *J Cogn Neurosci* 24 (8): 1664–1670.

Schlemm E, Cheng B, Fischer F, Hilgetag C, Gerloff C, et al. 2017. Altered topology of structural brain networks in patients with Gilles de la Tourette syndrome. *Sci Rep* 7 (1): 10606.

Schmitt O, and Eipert P. 2012. neuroVIISAS: Approaching multiscale simulation of the rat connectome. *Neuroinformatics* 10 (3): 243–267.

Schmitt O, Eipert P, Philipp K, Kettlitz R, Fuellen G, et al. 2012. The intrinsic connectome of the rat amygdala. *Front Neural Circuits* 6: 81.

Scholz J, Klein MC, Behrens TE, and Johansen-Berg H. 2009. Training induces changes in white-matter architecture. *Nat Neurosci* 12 (11): 1370–1371.

Schrepf A, Kaplan CM, Ichesco E, Larkin T, Harte SE, et al. 2018. A multi-modal MRI study of the central response to inflammation in rheumatoid arthritis. *Nat Commun* 9 (1): 2243.

Schumacher J, Peraza LR, Firbank M, Thomas AJ, Kaiser M, et al. 2019. Dysfunctional brain dynamics and their origin in Lewy body dementia. *Brain* 142 (6): 1767–1782.

Schwikowski B, Uetz P, and Fields S. 2000. A network of protein-protein interactions in yeast. *Nat Biotechnol* 18 (12): 1257–1261.

Scorcioni R, Polavaram S, and Ascoli GA. 2008. L-measure: A web-accessible tool for the analysis, comparison and search of digital reconstructions of neuronal morphologies. *Nat Protoc* 3: 866–876.

Segev R, Benveniste M, Shapira Y, and Ben-Jacob E. 2003. Formation of electrically active clusterized neural networks. *Phys Rev Lett* 90 (16): 168101.

Seguin C, van den Heuvel MP, and Zalesky A. 2018. Navigation of brain networks. *Proc Natl Acad Sci USA* 115 (24): 6297–6302.

Seldon HL. 2005. Does brain white matter growth expand the cortex like a balloon? Hypothesis and consequences. *Laterality* 10 (1): 81–95.

Senden M, Reuter N, van den Heuvel MP, Goebel R, Deco G, et al. 2018. Task-related effective connectivity reveals that the cortical rich club gates cortex-wide communication. *Hum Brain Mapp* 39 (3): 1246–1262.

Sernagor E, Eglen S, Harris B, and Wong R. 2006. *Retinal development*. Cambridge: Cambridge University Press.

Seung HS. 2009. Reading the book of memory: Sparse sampling versus dense mapping of connectomes. *Neuron* 62 (1): 17–29.

Shanahan M, Bingman VP, Shimizu T, Wild M, and Gunturkun O. 2013. Large-scale network organization in the avian forebrain: A connectivity matrix and theoretical analysis. *Front Comput Neurosci* 7: 89.

Shaw P, Kabani NJ, Lerch JP, Eckstrand K, Lenroot R, et al. 2008. Neurodevelopmental trajectories of the human cerebral cortex. *J Neurosci* 28 (14): 3586–3594.

Sherry DF, and Hoshooley JS. 2010. Seasonal hippocampal plasticity in food-storing birds. *Philos Trans R Soc Lond B Biol Sci* 365 (1542): 933–943.

Shih CT, Sporns O, Yuan SL, Su TS, Lin YJ, et al. 2015. Connectomics-based analysis of information flow in the *Drosophila* brain. *Curr Biol* 25 (10): 1249–1258.

Shimizu T, and Bowers AN. 1999. Visual pathways in the avian telencephalon: Evolutionary implications *Behav Brain Res* 98: 183–191.

Silva CG, Peyre E, and Nguyen L. 2019. Cell migration promotes dynamic cellular interactions to control cerebral cortex morphogenesis. *Nat Rev Neurosci* 20 (6): 318–329.

Sinha N, Dauwels J, Kaiser M, Cash SS, Brandon Westover M, et al. 2017. Predicting neurosurgical outcomes in focal epilepsy patients using computational modelling. *Brain* 140 (2): 319–332.

Sinha N, Wang Y, Dauwels J, Kaiser M, Thesen T, et al. 2019. Computer modelling of connectivity change suggests epileptogenesis mechanisms in idiopathic generalised epilepsy. *NeuroImage Clin* 21: 101655.

Skudlarski P, Jagannathan K, Anderson K, Stevens MC, Calhoun VD, et al. 2010. Brain connectivity is not only lower but different in schizophrenia: A combined anatomical and functional approach. *Biol Psychiatry* 68 (1): 61–69.

Sole-Casals J, Serra-Grabulosa JM, Romero-Garcia R, Vilaseca G, Adan A, et al. 2019. Structural brain network of gifted children has a more integrated and versatile topology. *Brain Struct Funct* 224 (7): 2373–2383.

Song H, and Poo M. 2001. The cell biology of neuronal navigation. *Nat Cell Biol* 3 (3): E81–88.

Sowell ER, Thompson PM, Leonard CM, Welcome SE, Kan E, et al. 2004. Longitudinal mapping of cortical thickness and brain growth in normal children. *J Neurosci* 24 (38): 8223–8231.

Spear PD, Tong L, and McCall MA. 1988. Functional influence of areas 17, 18, and 19 on lateral suprasylvian cortex in kittens and adult cats: Implications for compensation following early visual cortex damage. *Brain Res* 447 (1): 79–91.

References

Sperry RW. 1963. Chemoaffinity in the orderly growth of nerve fiber pattern and connections. *Proc Natl Acad Sci USA* 50: 703–710.

Sporns O. 2011. The non-random brain: Efficiency, economy, and complex dynamics. *Front Comput Neurosci* 5: 5.

Sporns O. 2013. Network attributes for segregation and integration in the human brain. *Curr Opin Neurobiol* 23 (2): 162–171.

Sporns O, and Betzel RF. 2016. Modular brain networks. *Annu Rev Psychol* 67: 613–640.

Sporns O, Chialvo DR, Kaiser M, and Hilgetag CC. 2004. Organization, development and function of complex brain networks. *Trends Cogn Sci* 8 (9): 418–425.

Sporns O, Honey CJ, and Kötter R. 2007. Identification and classification of hubs in brain networks. *PLOS One* 2 (10): e1049.

Sporns O, and Kötter R. 2004. Motifs in brain networks. *PLOS Biol* 2 (11): e369.

Sporns O, Tononi G, and Edelman GM. 2000. Theoretical neuroanatomy: Relating anatomical and functional connectivity in graphs and cortical connection matrices. *Cereb Cortex* 10 (2): 127–141.

Sporns O, Tononi G, and Kötter R. 2005. The human connectome: A structural description of the human brain. *PLOS Comput Biol* 1 (4): e42.

Stam CJ. 2014. Modern network science of neurological disorders. *Nat Rev Neurosci* 15 (10): 683–695.

Stam CJ, Hillebrand A, Wang H, and Van Mieghem P. 2010. Emergence of modular structure in a large-scale brain network with interactions between dynamics and connectivity. *Front Comput Neurosci* 4 (133).

Stam CJ, Jones BF, Nolte G, Breakspear M, and Scheltens P. 2007. Small-world networks and functional connectivity in Alzheimer's disease. *Cereb Cortex* 17 (1): 92–99.

Stauffer D, and Aharony A. 2003. *Introduction to percolation theory*. Rev. 2nd ed. London: Routledge.

Stepanyants A, Hof PR, and Chklovskii DB. 2002. Geometry and structural plasticity of synaptic connectivity. *Neuron* 34 (2): 275–288.

Stephan KE. 2013. The history of CoCoMac. *Neuroimage* 80: 46–52.

Stephan KE, Baldeweg T, and Friston KJ. 2006. Synaptic plasticity and dysconnection in schizophrenia. *Biol Psychiatry* 59 (10): 929–939.

Stephan KE, Kamper L, Bozkurt A, Burns GA, Young MP, et al. 2001. Advanced database methodology for the collation of connectivity data on the macaque brain (CoCoMac). *Philos Trans R Soc Lond B Biol Sci* 356 (1412): 1159–1186.

Sterratt D, Graham B, Gillies A, and Willshaw D. 2011. *Principles of computational modelling in neuroscience*. Cambridge: Cambridge University Press.

Stief F, Zuschratter W, Hartmann K, Schmitz D, and Draguhn A. 2007. Enhanced synaptic excitation-inhibition ratio in hippocampal interneurons of rats with temporal lobe epilepsy. *Eur J Neurosci* 25 (2): 519–528.

Striedter GF. 2005. *Principles of brain evolution*. Sunderland, MA: Sinauer Associates.

Stumpf MPH, Wiuf C, and May RM. 2005. Subnets of scale-free networks are not scale-free: Sampling properties of networks. *Proc Natl Acad Sci USA* 102: 4221–4224.

Sukhinin DI, Engel AK, Manger P, and Hilgetag CC. 2016. Building the Ferretome. *Front Neuroinform* 10: 16.

Sulston JE, and Horvitz HR. 1977. Post-embryonic cell lineages of the nematode, *Caenorhabditis elegans*. *Dev Biol* 56 (1): 110–156.

Sulston JE, Schierenberg E, White JG, and Thomson JN. 1983. The embryonic cell lineage of the nematode *Caenorhabditis elegans*. *Dev Biol* 100 (1): 64–119.

Sun Y, Chen Y, Lee R, Bezerianos A, Collinson SL, et al. 2016. Disruption of brain anatomical networks in schizophrenia: A longitudinal, diffusion tensor imaging based study. *Schizophr Res* 171 (1–3): 149–157.

Sundaram SK, Kumar A, Makki MI, Behen ME, Chugani HT, et al. 2008. Diffusion tensor imaging of frontal lobe in autism spectrum disorder. *Cereb Cortex* 18 (11): 2659–2665.

Sunkin SM, Ng L, Lau C, Dolbeare T, Gilbert TL, et al. 2013. Allen brain atlas: An integrated spatiotemporal portal for exploring the central nervous system. *Nucleic Acids Res* 41: D996–D1008.

Supekar K, Musen M, and Menon V. 2009. Development of large-scale functional brain networks in children. *PLOS Biol* 7 (7): e1000157.

Swanson LW. 2018. *Brain maps 4.0—structure of the rat brain*: An open access atlas with global nervous system nomenclature ontology and flatmaps. *J Comp Neurol* 526 (6): 935–943.

Swanson LW, Hahn JD, Jeub LGS, Fortunato S, and Sporns O. 2018. Subsystem organization of axonal connections within and between the right and left cerebral cortex and cerebral nuclei (endbrain). *Proc Natl Acad Sci USA* 115 (29): E6910–E6919.

Swanson LW, Sporns O, and Hahn JD. 2016. Network architecture of the cerebral nuclei (basal ganglia) association and commissural connectome. *Proc Natl Acad Sci USA* 113 (40): E5972–E5981.

Swanson LW, Sporns O, and Hahn JD. 2019. The network organization of rat intrathalamic macroconnections and a comparison with other forebrain divisions. *Proc Natl Acad Sci USA* 116 (27): 13661–13669.

Swindale NV. 1980. A model for the formation of ocular dominance stripes. *Proc R Soc Lond B Biol Sci* 208: 243–264.

References

Takahashi E, Dai G, Wang R, Ohki K, Rosen GD, et al. 2010. Development of cerebral fiber pathways in cats revealed by diffusion spectrum imaging. *Neuroimage* 49 (2): 1231–1240.

Tallinen T, Chung JY, Biggins JS, and Mahadevan L. 2014. Gyrification from constrained cortical expansion. *Proc Natl Acad Sci USA* 111 (35): 12667–12672.

Tallinen T, Chung JY, Rousseau F, Girard N, Lefevre J, et al. 2016. On the growth and form of cortical convolutions. *Nature Phys* 12 (6): 588–593.

Tamnes CK, Ostby Y, Fjell AM, Westlye LT, Due-Tonnessen P, et al. 2010. Brain maturation in adolescence and young adulthood: Regional age-related changes in cortical thickness and white matter volume and microstructure. *Cereb Cortex* 20 (3): 534–548.

Taylor PN, Han CE, Schoene-Bake JC, Weber B, and Kaiser M. 2015. Structural connectivity changes in temporal lobe epilepsy: Spatial features contribute more than topological measures. *Neuroimage Clin* 8 (0): 322–328.

Taylor PN, Thomas J, Sinha N, Dauwels J, Kaiser M, et al. 2015. Optimal control based seizure abatement using patient derived connectivity. *Front Neurosci* 9: 202.

Tecchio F, Cottone C, Porcaro C, Cancelli A, Di Lazzaro V, et al. 2018. Brain functional connectivity changes after transcranial direct current stimulation in epileptic patients. *Front Neural Circuits* 12: 44.

Thanarajah SE, Han CE, Rotarska-Jagiela A, Singer W, Deichmann R, et al. 2016. Abnormal connectional fingerprint in schizophrenia: A novel network analysis of diffusion tensor imaging data. *Front Psychiatry* 7: 114.

Thiebaut de Schotten M, Cohen L, Amemiya E, Braga LW, and Dehaene S. 2014. Learning to read improves the structure of the arcuate fasciculus. *Cereb Cortex* 24 (4): 989–995.

Thornton C, Hutchings F, and Kaiser M. 2019. The Virtual Electrode Recording Tool for EXtracellular potentials (VERTEX) Version 2.0: Modelling *in vitro* electrical stimulation of brain tissue. *Wellcome Open Res* 4: 20.

Tomasi S, Caminiti R, and Innocenti GM. 2012. Areal differences in diameter and length of corticofugal projections. *Cereb Cortex* 22 (6): 1463–1472.

Tomassy GS, Berger DR, Chen HH, Kasthuri N, Hayworth KJ, et al. 2014. Distinct profiles of myelin distribution along single axons of pyramidal neurons in the neocortex. *Science* 344 (6181): 319–324.

Tomassy GS, Dershowitz LB, and Arlotta P. 2016. Diversity matters: A revised guide to myelination. *Trends Cell Biol* 26 (2): 135–147.

Tomsett RJ, Ainsworth M, Thiele A, Sanayei M, Chen X, et al. 2015. Virtual Electrode Recording Tool for EXtracellular potentials (VERTEX): Comparing multi-electrode recordings from simulated and biological mammalian cortical tissue. *Brain Struct Funct* 220 (4): 2333–2353.

Tononi G, Sporns O, and Edelman GM. 1999. Measures of degeneracy and redundancy in biological networks. *Proc Natl Acad Sci USA* 96 (6): 3257–3262.

Toro R, and Burnod Y. 2005. A morphogenetic model for the development of cortical convolutions. *Cereb Cortex* 15 (12): 1900–1913.

Toro R, Perron M, Pike B, Richer L, Veillette S, et al. 2008. Brain size and folding of the human cerebral cortex. *Cereb Cortex* 18 (10): 2352–2357.

Towlson EK, Vertes PE, Ahnert SE, Schafer WR, and Bullmore ET. 2013. The rich club of the *C. elegans* neuronal connectome. *J Neurosci* 33 (15): 6380–6387.

Trappenberg TP. 2010. *Fundamentals of computational neuroscience*. Oxford: Oxford University Press.

Triplett MA, Avitan L, and Goodhill GJ. 2018. Emergence of spontaneous assembly activity in developing neural networks without afferent input. *PLOS Comput Biol* 14 (9): e1006421.

Tschentscher N, Mitchell D, and Duncan J. 2017. Fluid intelligence predicts novel rule implementation in a distributed frontoparietal control network. *J Neurosci* 37 (18): 4841–4847.

Tuch DS, Reese TG, Wiegell MR, and Wedeen VJ. 2003. Diffusion MRI of complex neural architecture. *Neuron* 40 (5): 885–895.

Turing AM. 1952. The chemical basis of morphogenesis. *Philos Trans R Soc Lond B Biol Sci* 237 (641): 37–72.

Tymofiyeva O, Hess CP, Ziv E, Lee PN, Glass HC, et al. 2013. A DTI-based template-free cortical connectome study of brain maturation. *PLOS One* 8 (5): e63310.

Utton MA, Noble WJ, Hill JE, Anderton BH, and Hanger DP. 2005. Molecular motors implicated in the axonal transport of tau and alpha-synuclein. *J Cell Sci* 118 (Pt 20): 4645–4654.

Vaessen MJ, Braakman HM, Heerink JS, Jansen JF, Debeij-van Hall MH, et al. 2013. Abnormal modular organization of functional networks in cognitively impaired children with frontal lobe epilepsy. *Cereb Cortex* 23 (8): 1997–2006.

van den Heuvel MI, Turk E, Manning JH, Hect J, Hernandez-Andrade E, et al. 2018. Hubs in the human fetal brain network. *Dev Cogn Neurosci* 30: 108–115.

van den Heuvel MP, Bullmore ET, and Sporns O. 2016. Comparative connectomics. *Trends Cogn Sci* 20: 345–361.

van den Heuvel MP, Kersbergen KJ, de Reus MA, Keunen K, Kahn RS, et al. 2015. The neonatal connectome during preterm brain development. *Cereb Cortex* 25 (9): 3000–3013.

van den Heuvel MP, Mandl RC, Stam CJ, Kahn RS, and Hulshoff Pol HE. 2010. Aberrant frontal and temporal complex network structure in schizophrenia: A graph theoretical analysis. *J Neurosci* 30 (47): 15915–15926.

References

van den Heuvel MP, and Sporns O. 2011. Rich-club organization of the human connectome. *J Neurosci* 31 (44): 15775–15786.

van den Heuvel MP, Stam CJ, Kahn RS, and Hulshoff Pol HE. 2009. Efficiency of functional brain networks and intellectual performance. *J Neurosci* 29 (23): 7619–7624.

Van Essen DC. 1997. A tension-based theory of morphogenesis and compact wiring in the central nervous system. *Nature* 385 (6614): 313–318.

Van Essen DC, and Drury HA. 1997. Structural and functional analyses of human cerebral cortex using a surface-based atlas. *J Neurosci* 17 (18): 7079–7102.

Van Essen DC, Ugurbil K, Auerbach E, Barch D, Behrens TE, et al. 2012. The Human Connectome Project: A data acquisition perspective. *Neuroimage* 62 (4): 2222–2231.

van Ooyen A. 2001. Competition in the development of nerve connections: A review of models. *Network* 12 (1): R1–47.

van Ooyen A. 2003. *Modeling neural development*. Cambridge: MIT Press.

van Ooyen A. 2011. Using theoretical models to analyse neural development. *Nat Rev Neurosci* 12 (6): 311–326.

van Ooyen A, and Butz-Ostendorf M. 2017. *The rewiring brain: A computational approach to structural plasticity in the adult brain*. London: Academic Press.

van Ooyen A, Carnell A, de Ridder S, Tarigan B, Mansvelder HD, et al. 2014. Independently outgrowing neurons and geometry-based synapse formation produce networks with realistic synaptic connectivity. *PLOS One* 9 (1): e85858.

van Ooyen A, and Willshaw DJ. 1999. Competition for neurotrophic factor in the development of nerve connections. *Proc Biol Sci* 266 (1422): 883–892.

Varier S, and Kaiser M. 2011. Neural development features: Spatio-temporal development of the *Caenorhabditis elegans* neuronal network. *PLOS Comput Biol* 7 (1): e1001044.

Varier S, Kaiser M, and Forsyth R. 2011. Establishing, versus maintaining, brain function: A neuro-computational model of cortical reorganization after injury to the immature brain. *J Int Neuropsychol Soc* 17 (6): 1030–1038.

Varshney LR, Chen BL, Paniagua E, Hall DH, and Chklovskii DB. 2011. Structural properties of the *Caenorhabditis elegans* neuronal network. *PLOS Comput Biol* 7 (2): e1001066.

Váša F, Romero-Garcia R, Kitzbichler MG, Seidlitz J, Whitaker KJ, et al. 2019. Conservative and disruptive modes of adolescent change in brain functional connectivity. *bioRxiv*: 604843.

Vázquez A, Flammini A, Maritan A, and Vespignani A. 2003. Modeling of protein interaction networks. *Complexus* 1 (1): 38–44.

Verhagen L, Gallea C, Folloni D, Constans C, Jensen DE, et al. 2019. Offline impact of transcranial focused ultrasound on cortical activation in primates. *Elife* 8.

Voets NL, Bernhardt BC, Kim H, Yoon U, and Bernasconi N. 2011. Increased temporolimbic cortical folding complexity in temporal lobe epilepsy. *Neurology* 76 (2): 138–144.

von der Malsburg C. 1995. Binding in models of perception and brain function. *Curr Opin Neurobiol* 5 (4): 520–526.

von der Malsburg C, and Willshaw DJ. 1976. A mechanism for producing continuous neural mappings: Ocularity dominance stripes and ordered retino-tectal projections. *Exp Brain Res* no. 1: 463–469.

von Neumann J. 1958. *The computer and the brain*. New Haven: Yale University Press.

Wagner A. 2000. Robustness against mutations in genetic networks of yeast. *Nat Genet* 24 (4): 355–361.

Wang Y, Hutchings F, and Kaiser M. 2015. Computational modeling of neurostimulation in brain diseases. *Prog Brain Res* 222: 191–228.

Wang Y, Necus J, Kaiser M, and Mota B. 2016. Universality in human cortical folding in health and disease. *Proc Natl Acad Sci USA* 113 (45): 12820–12825.

Wang Y, Necus J, Rodriguez LP, Taylor PN, and Mota B. 2019. Human cortical folding across regions within individual brains follows universal scaling law. *Commun Biol* 2: 191.

Wang Y, Trevelyan AJ, Valentin A, Alarcon G, Taylor PN, et al. 2017. Mechanisms underlying different onset patterns of focal seizures. *PLOS Comput Biol* 13 (5): e1005475.

Wasserman S, and Faust K. 1994. *Social network analysis: Methods and applications: Vol. 8. Structural analysis in the social sciences*. Cambridge: Cambridge University Press.

Watanabe T, and Rees G. 2015. Age-associated changes in rich-club organisation in autistic and neurotypical human brains. *Sci Rep* 5: 16152.

Watts DJ. 1999. *Small worlds*. Princeton: Princeton University Press.

Watts DJ, and Strogatz SH. 1998. Collective dynamics of 'small-world' networks. *Nature* 393 (6684): 440–442.

Weickenmeier J, Kuhl E, and Goriely A. 2018. Multiphysics of prionlike diseases: Progression and atrophy. *Phys Rev Lett* 121 (15): 158101.

Welker W. 1990. "Why does cerebral cortex fissure and fold?" In *Cerebral cortex*, edited by Jones, EG and Peters, A, 3–136. Boston: Springer US.

Wen H, Liu Y, Wang S, Zhang J, Peng Y, et al. 2017. "Diffusion tractography and graph theory analysis reveal the disrupted rich-club organization of white matter structural networks in early Tourette syndrome children." Proceedings Vol. 10137. *SPIE Medical Imaging*. doi:10.1117/12.2254093.

Wen Q, and Chklovskii DB. 2005. Segregation of the brain into gray and white matter: A design minimizing conduction delays. *PLOS Comput Biol* 1 (7): e78.

Wen Q, and Chklovskii DB. 2008. A cost-benefit analysis of neuronal morphology. *J Neurophysiol* 99 (5): 2320–2328.

White JG, Southgate E, Thomson JN, and Brenner S. 1986. The structure of the nervous system of the nematode *Caenorhabditis elegans*. *Philos Trans R Soc Lond B Biol Sci* 314 (1165): 1–340.

White T, Su S, Schmidt M, Kao CY, and Sapiro G. 2010. The development of gyrification in childhood and adolescence. *Brain Cogn* 72 (1): 36–45.

Widge AS, Zorowitz S, Basu I, Paulk AC, Cash SS, et al. 2019. Deep brain stimulation of the internal capsule enhances human cognitive control and prefrontal cortex function. *Nat Commun* 10 (1): 1536.

Wiecki TV, and Frank MJ. 2010. Neurocomputational models of motor and cognitive deficits in Parkinson's disease. In *Progress in brain research: Vol. 183. Recent advances in Parkinson's disease: Basic research*, 275–297. Amsterdam: Elsevier.

Wiesel TN. 1982. Postnatal-development of the visual-cortex and the influence of environment. *Nature* 299 (5884): 583–591.

Williams DL, and Minshew NJ. 2007. Understanding autism and related disorders: What has imaging taught us? *Neuroimaging Clin N Am* 17 (4): 495–509.

Willshaw DJ. 1981. The establishment and the subsequent eliminiation of polyneural innervation of developing muscle: Theoretical considerations. *Proc R Soc Lond B Biol Sci* 212 (1187): 233–252.

Willshaw DJ, and von der Malsburg C. 1976. How patterned neural connections can be set up by self-organization. *Proc R Soc Lond B Biol Sci* 194 (1117): 431–445.

Wolf L, Goldberg C, Manor N, Sharan R, and Ruppin E. 2011. Gene expression in the rodent brain is associated with its regional connectivity. *PLOS Comput Biol* 7 (5): e1002040.

Woodward ND, Rogers B, and Heckers S. 2011. Functional resting-state networks are differentially affected in schizophrenia. *Schizophr Res* 130 (1–3): 86–93.

Xie L, Kang H, Xu Q, Chen MJ, Liao Y, et al. 2013. Sleep drives metabolite clearance from the adult brain. *Science* 342: 373–377.

Xu G, Knutsen AK, Dikranian K, Kroenke CD, Bayly PV, et al. 2010. Axons pull on the brain, but tension does not drive cortical folding. *J Biomech Eng* 132 (7): 071013.

Yamamoto N, Tamada A, and Murakami F. 2002. Wiring of the brain by a range of guidance cues. *Prog Neurobiol* 68 (6): 393–407.

Yan G, Vertes PE, Towlson EK, Chew YL, Walker DS, et al. 2017. Network control principles predict neuron function in the *Caenorhabditis elegans* connectome. *Nature* 550 (7677): 519–523.

Yang Z, Xu Y, Xu T, Hoy CW, Handwerker DA, et al. 2014. Brain network informed subject community detection in early-onset schizophrenia. *Sci Rep* 4: 5549.

Yeh FC, Badre D, and Verstynen T. 2016. Connectometry: A statistical approach harnessing the analytical potential of the local connectome. *Neuroimage* 125: 162–171.

Yizhar O, Fenno LE, Prigge M, Schneider F, Davidson TJ, et al. 2011. Neocortical excitation/inhibition balance in information processing and social dysfunction. *Nature* 477 (7363): 171–178.

Yoo SW, Han CE, Shin JS, Seo SW, Na DL, et al. 2015. A network flow-based analysis of cognitive reserve in normal ageing and Alzheimer's disease. *Sci Rep* 5: 10057.

Yoshiyama Y, Lee VMY, and Trojanowski JQ. 2013. Therapeutic strategies for tau mediated neurodegeneration. *J Neurol Neurosurg Psychiatry* 84 (7): 784–795.

Young MP. 1992. Objective analysis of the topological organization of the primate cortical visual system. *Nature* 358 (6382): 152–155.

Young MP. 1993. The organization of neural systems in the primate cerebral cortex. *Proc Biol Sci* 252 (1333): 13–18.

Young MP. 2000. The architecture of visual cortex and inferential processes in vision. *Spat Vis* 13 (2–3): 137–146.

Young MP, and Scannell JW. 1996. Component-placement optimization in the brain. *Trends Neurosci* 19 (10): 413–415.

Ypma RJ, and Bullmore ET. 2016. Statistical analysis of tract-tracing experiments demonstrates a dense, complex cortical network in the mouse. *PLOS Comput Biol* 12 (9): e1005104.

Yu YC, Bultje RS, Wang X, and Shi SH. 2009. Specific synapses develop preferentially among sister excitatory neurons in the neocortex. *Nature* 458 (7237): 501–504.

Yu YC, He S, Chen S, Fu Y, Brown KN, et al. 2012. Preferential electrical coupling regulates neocortical lineage-dependent microcircuit assembly. *Nature* 486 (7401): 113–117.

Zalesky A, Fornito A, Seal ML, Cocchi L, Westin CF, et al. 2011. Disrupted axonal fiber connectivity in schizophrenia. *Biol Psychiatry* 69 (1): 80–89.

Zamora-Lopez G, Zhou C, and Kurths J. 2010. Cortical hubs form a module for multisensory integration on top of the hierarchy of cortical networks. *Front Neuroinform* 4: 1.

Zhan Y, Paolicelli RC, Sforazzini F, Weinhard L, Bolasco G, et al. 2014. Deficient neuron-microglia signaling results in impaired functional brain connectivity and social behavior. *Nat Neurosci* 17 (3): 400–406.

Zhang Z, Liao W, Chen H, Mantini D, Ding JR, et al. 2011. Altered functional-structural coupling of large-scale brain networks in idiopathic generalized epilepsy. *Brain* 134 (Pt 10): 2912–2928.

Zhao T, Mishra V, Jeon T, Ouyang M, Peng Q, et al. 2019. Structural network maturation of the preterm human brain. *Neuroimage* 185: 699–710.

Zhao T, Xu Y, and He Y. 2019. Graph theoretical modeling of baby brain networks. *Neuroimage* 185: 711–727.

References

Zhou C, Zemanova L, Zamora G, Hilgetag CC, and Kurths J. 2006. Hierarchical organization unveiled by functional connectivity in complex brain networks. *Phys Rev Lett* 97 (23): 238103.

Zhou C, Zemanová L, Zamora-Lopez G, Hilgetag CC, and Kurths J. 2007. Structure–function relationship in complex brain networks expressed by hierarchical synchronization. *New J Phys* 9: 178.

Zhou S, and Mondragon RJ. 2004. The rich-club phenomenon in the internet topology. *IEEE Commun Lett* 8 (3): 180–182.

Zhou YD, Lee S, Jin Z, Wright M, Smith SE, et al. 2009. Arrested maturation of excitatory synapses in autosomal dominant lateral temporal lobe epilepsy. *Nat Med* 15 (10): 1208–1214.

Zhu F, Cizeron M, Qiu Z, Benavides-Piccione R, Kopanitsa MV, et al. 2018. Architecture of the mouse brain synaptome. *Neuron* 99 (4): 781–799.e10.

Zielinski BA, Prigge MB, Nielsen JA, Froehlich AL, Abildskov TJ, et al. 2014. Longitudinal changes in cortical thickness in autism and typical development. *Brain* 137 (Pt 6): 1799–1812.

Zingg B, Hintiryan H, Gou L, Song MY, Bay M, et al. 2014. Neural networks of the mouse neocortex. *Cell* 156 (5): 1096–1111.

Ziv NE, and Brenner N. 2018. Synaptic tenacity or lack thereof: Spontaneous remodeling of synapses. *Trends Neurosci* 41 (2): 89–99.

Zubler F, and Douglas R. 2009. A framework for modeling the growth and development of neurons and networks. *Front Comput Neurosci* 3: 25.

Zubler F, Hauri A, Pfister S, Bauer R, Anderson JC, et al. 2013. Simulating cortical development as a self constructing process: A novel multi-scale approach combining molecular and physical aspects. *PLOS Comput Biol* 9 (8): e1003173.

Zubler F, Hauri A, Pfister S, Whatley AM, Cook M, et al. 2011. An instruction language for self-construction in the context of neural networks. *Front Comput Neurosci* 5: 57.

Zuo XN, He Y, Betzel RF, Colcombe S, Sporns O, et al. 2017. Human connectomics across the life span. *Trends Cogn Sci* 21 (1): 32–45.

Index

Abbott, L. F., 147
Acetylcholine, 190
Achacoso, T. B., 7, 38
Achard, S, 15–16, 22, 25
Actin, 138
Addington, A. M., 33
Adhesion, 95, 97–98, 161
Adolescence, 129, 134, 145–146, 156–157
Aerts, H., 170
Aggregation, 38
Agnosia, 154
Agranular, 82, 109
Ahn, Y. Y., 18
Akil, H., 9
Albert, R., 103, 111
Alexander-Bloch, A. F., 156
Alexia, 154
Alfaro-Almagro, F., 63
Allocortex, 79–80
Alstott, J., 179
Altman, J., 120
Altun, Z. F., 44, 123
Alzheimer's disease
  in cats and dogs, 58
  as a type of dementia, 165
  histological changes, 166
  and limbic areas, 82
  modeling disease progression, 172–176
  and neuronal activity, 197
Amaral, L. A. N., 18, 108
Amari, S., 77

Amniotes, 83
Amygdala, 51, 169, 176, 184
Amyloid proteins, 58, 148, 166–168, 170–172, 174
Amyotrophic Lateral Sclerosis (ALS), 168
Anagnostou, E., 157
Andersen, S. L., 120
Anesthesia, 147
Annelids, 38, 45
Anorexia, 135
Aparício, D., 20
Aphasia, 154, 192
Apoptosis
  cortical folding, 132, 134
  layer formation, 81, 86, 88
  module formation, 129
  overview, 67, 81
  time windows, 120
Archicortex, 79, 113
Armstrong, E., 146
Arnatkeviciute, A., 118
Arthropods, 39, 44, 47
Ascoli, G. A., 51, 89
Asllani, M., 196
Asperger syndrome, 156
Astrocytes, 93
Ataxia, 83
Attention deficit hyperactivity disorder (ADHD), 135, 162
Autism, 80, 135, 153, 156–157, 160–162

Axons
　fasciculation, 89, 98
　growth, 92, 95–99, 102
　pruning, 102, 144
　sprouting, 180

Bacteria, 113, 178
Bakola, S, 60
Ball, G., 157
Barabasi, A. L., 103, 111
Barbas, H., 26, 133–134
Barbeito-Andres, J., 110, 162
Barone, P., 120
Baruch, L., 92
Basket cell, 89, 195
Bassett, D., 155, 163
Bathelt, J., 151
Bauer, R.
　hub formation, 108, 111–114, 117
　layer formation, 80–81, 86
　patchy connectivity, 93, 104–105
Bayer, S. A., 120
Bayly, P. V., 133
Beaulieu, C., 143
Belfore, L. A., 167
Belmonte, M. K., 157
Bengtsson, S. L., 94
Bernardoni, F., 135
Bestmann, S., 193
Betweenness
　centrality, 14
　and hubs, 107
　in humans, 60
　and lesions, 179
　in pigeons, 48
　in rats, 55
Betzel, R. F., 94, 129, 145, 159
Beul, S. F., 56
Bhagwat, N., 174
Bienkowski, M. S., 50
Bilingual, 94
Binzegger, T., 25, 56, 92
Blaxter, M., 44

Boksa, P., 155
Bollobas, B., 22
Bonifazi, P., 22
Bonilha, L., 158
Borisyuk, R., 43, 91, 95
Bota, M., 51
Boutons, 93, 181, 183
Bowers, A. N., 48
Bozzi, Y., 80, 158, 161
Braak, H., 167
Bradykinesia, 165
Brain dynamics, 35, 170
Braitenberg, V., 92, 102
Brandes, U., 14
Breakspear, M., 25
Brenner, S., 148
Briscoe, S. D., 83
Broadcasters, 12, 46, 151
Brodmann, K., 68, 76, 79
Bullmore, E. T., 15–16, 50, 109, 153, 158
Bundy, D. T., 180–181, 191
Burnod, Y., 71
Buschke, H., 165
Butz, M., 89, 181–183, 186, 191

Cabral, J., 35
Caenorhabditis elegans
　cell death, 81
　connectivity, 15, 19, 24, 37–38, 43–44
　gene expression, 50, 92
　hubs, 112
　network development, 98, 115, 120–124
　spatial network organization, 26–31, 98
　stimulation, 196
Caffrey, J. R., 85
Cahalane, D. J., 84–85
Caleo, M., 161
Callithrix jacchus. See Marmoset monkey
Cancer, 86
Cao, M., 143
Carnivores, 55, 121
Castellanos, N. P., 181

Cat
  axonal branching, 99
  daisy architecture, 93
  hub and rich-club organization, 56, 109, 112
  motifs, 19
  network robustness, 23–24, 178–180
  small-world organization, 40, 55–58
  spatial growth, 102–104
Catani, M., 154
Cell lineage
  C. elegans, 43–44, 81
  drosophila, 98
  mice neocortex, 123
  module formation, 119–123, 125
Centrality. See also Betweenness
  and hubs, 107
  and lesions, 179
  betweenness, 14, 48, 55, 60, 107, 179
  nodes and edge, 14
Cephalization, 38–39
Cephalopods, 37, 39
Cerebellum, 5, 70, 89, 110, 150
Cerebrospinal, 147, 171, 176
Chalmers, K., 99–100
Channelrhodopsin, 194
Characteristic path length. See also Efficiency; Small-world networks
  brain diseases and, 156, 160, 169
  and hubs, 109
  network robustness, 178–179
  small-world organization, 15–16, 24
  wiring optimization, 28–31
Chaudhuri, D., 98
Chavez, M., 158
Chechik, G., 144
Chemoaffinity hypothesis, 76, 78
Chemoreceptors, 38
Chen, B. L., 123
Chen, H., 133
Chen, X., 197
Cherniak, C., 28, 44
Chiang, A. S., 45, 121
Childhood, 143, 145, 156–157

Chimpanzee, 58
Chklovskii, D. B., 28, 31
Choe, Y., 43, 123
Chordates, 39
Chronoarchitecture, 120
Cingulate cortex, 48–49, 109, 157, 159, 167
Clauset, A., 16, 22
Clustering coefficient. See also Efficiency; Small-world networks
  brain diseases and, 156–157, 162, 169
  during development, 145
  local and global, 12–13, 15–16
  small-world organization, 24, 129
  spatial growth, 103–104
Clusters. See Modules
Cnidaria, 37–38
CoCoMac, 58
Coevolution, 151
Cohen Kadosh, R., 198
Colizza, V., 110
Colliculus, 77–78, 183
Collin, G., 68, 110, 156
Collins, C. E., 86
Columba livia, 48. See also Pigeon
Column
  cortical, 25, 84, 119, 195
  ocular dominance, 65, 70, 74–75, 121, 161
Commissures, 39
Competition
  activity-dependent, 74–77
  for space, 99–102, 122
  maximizing fitness, 41
  molecular, 72, 74, 89
Component placement optimization (CPO), 28
Connectivity. See Axons; Effective connectivity; Functional connectivity; Structural connectivity
Connectome, 7–8, 25, 60
Connectometry, 94
Consciousness, 157
Contreras, J. A., 169
Cook, S. J., 37

Cope, T. E., 170
Cormen, T. H., 28
Corpus callosum, 147, 155, 157, 162
Cortex. *See* Cortical layers; Gray matter
Cortical folding, 131–136, 138, 140
Cortical layers, 82–83
Corticogenesis, 120
Costa, L. da F., 7, 32, 38, 90, 94
Coupe, P., 169
Courchesne, E., 156, 160
Cowan, W. M., 132, 144
Crawley, J. N., 80
Crepel, F., 70
Crick, F., 60
Critical period, 31, 121, 158
Csernansky, J. G., 135
Cycle, 8, 47
Cyron, D., 198
Cytoarchitecture, 55–56, 79, 82, 155

Daisy architecture, 93, 104
Damicelli, F., 129
Dancause, N., 180
Dayan, P., 147
Deafferentation, 181
Deafness, 154
Deco, G., 35
DeFelipe, J., 33
Degeneracy, 33, 177
Degree distribution, 20–24, 103–104, 107, 111, 116, 127
Deguchi, Y., 121
de Haan, W., 174, 197
Dehaene, S., 147
Delli Pizzi, S., 169
Dementia. *See* Alzheimer's disease; Lewy body dementia; Parkinson's disease
Demirtas, M., 158
Dendrites, 39, 86, 89, 99–102, 122–123, 144–145
Denk, W., 7
Depression, 135, 158, 162–163, 192
de Reus, M. A., 56

DeSalvo, M. N., 158
Deutocerebrum, 39
Developing Human Connectome Project, 60, 67, 143
Diesmann, M., 53
Diestel, R., 8
Diffusion Spectrum Imaging (DSI), 10–11, 18
Diffusion Tensor Imaging (DTI), 10–11, 27
Di Martino, A., 68, 163
Disconnection, 154, 161
Disconnexion, 154
Diseases. *See* Agnosia; Alzheimer's disease; Amyotrophic lateral sclerosis (ALS); Anorexia; Aphasia; Ataxia; Attention deficit hyperactivity disorder (ADHD); Autism; Deafness; Depression; Dyslexia; Epilepsy; Lewy body dementia; Parkinson's disease; Schizophrenia; Tourette syndrome; Williams syndrome
Dispersion, 159–160
Dopamine, 167, 197
Douglas, R. J., 82, 86, 93
Drosophila melanogaster, 44–47, 98, 121, 127, 129. *See also* Fruit fly
Druckmann, S, 121
Drury, H. A., 7, 131, 136–137
Dubois, B., 165
Duffing oscillator, 197
Dumas de la Roque, A., 160
Duncan, J., 151
Durbin, R. M., 37, 43–44, 98
Dysconnection, 155
Dyslexia, 135
Dysplasia, 110, 158
Dystonia, 189

Ebbesson, S. O. E., 114
Economy. *See* Wiring optimization
Edelman, G. M., 33
Edge density, 5, 14–16, 21, 24, 40, 55, 103–104
Effective connectivity, 7–8, 12, 110, 115

Efficiency. *See also* Characteristic path length; Clustering coefficient; Small-world networks
  brain diseases and, 156–158, 169, 183
  global, 15–16, 149, 183
  local, 16, 149
Eglen, S. J., 65, 71–72
Eguiluz, V. M., 22
Eickhoff, S. B., 5, 10
Eipert, P., 53
Electrocorticogram (ECoG), 190
Electroencephalography (EEG), 33, 167, 169–170, 176, 191
Emotion, 154–156
Ephrin, 76
Epigenetic, 34, 41
Epilepsy
  brain stimulation, 189–191, 196
  connectome changes, 157–158, 160
  hyperexcitability, 161
  pathological layer formation and cortical folding, 80, 83, 135
  surgery, 163, 179
Ercsey-Ravasz, M., 58, 95
Erdős, P., 21, 23–24, 116
Erlebach, T., 14
Euler, L., 8
Eutelic, 43
Evolution
  brain network, 37, 39
  cortical layers, 83
  human brain, 40
  parcellation of the cortex, 114
  protein-protein interaction networks, 115
  scaling laws, 138
Excitability, 157–158, 161, 185, 187, 190, 197

Fair, D. A., 129, 150
Fard, P. K., 99
Fasciculation, 89, 98
Fauth, M. J., 147
Felis silvestris catus. *See* Cat

Felleman, D. J., 7, 37, 40, 58
Ferret, 55
Ferretome, 55
ffytche, D. H., 154
Fields, R. D., 93
Filopodia, 95, 102
Finger, S., 177, 186
Fitzgibbon, S. P., 60, 67, 143
Fletcher, P. C., 133–135
Formatio reticularis, 39
Fornito, A., 72–73, 109, 155
Foutz, T. J., 194
Fractal, 56, 137
Franze, K., 95
Fratiglioni, L., 166
Freeman, D. C., 68–69
Fries, P., 151
Friston, K. J., 33, 154, 161, 177
Frontolimbic, 45, 56–57
Frontoparietal, 157, 169
Frontopolar, 191
Frontotemporal, 162, 166, 173, 192
Fruit fly, 43–45, 47, 121
Functional compensation, 177, 183–186
Functional connectivity
  brain diseases and, 155, 157–158, 161, 167, 169–172, 174
  brain stimulation and, 190–191, 197
  cognition and, 151, 153
  during development, 144–145
  hubs and, 110, 145
  lesions and, 180, 182
  reconstruction, 8–11
  scale-free networks and, 22
  spatial features, 95
  across species, 60
Functional magnetic resonance imaging (fMRI), 22, 27, 33, 63, 167, 170–171

Gaig, C., 165
Gallos, L. K., 45
Galton–Watson model, 99
Galvan, C. D., 161

Ganglia
  basal, 170, 197
  in C. elegans, 44, 81
  evolution of, 38–39
  in vertebrates, 40
Gao, W., 129
Garcia, K. E., 134, 137
Garcia-Cabezas, M. A., 82
Gastrulation, 67
Gender, 134
Genome, 33, 104, 178
Georgiadis, K., 175
Geschwind, N., 154
Geser, F., 165
Gewaltig, M.-O., 53
Gibbons, 58
Gibson, E. M., 94
Giedd, J. N., 143
Gierer, A., 71, 105
Gilbert, C. D., 93
Girvan, M., 16
Glasser, M. F., 5, 60
Gleissner, U., 179
Glial cells, 67, 132
Glioma, 86
Godfrey, K. B., 77
Gohlke, J. M., 84
Goldsworthy, M. R., 192
Goodall, S., 185
Goodfellow, M., 163
Goodhill, G. J., 92, 97
Görner, M., 25, 150
Goslin, K., 89
Goulas, A., 45, 127–128
Gouzé, J. L., 74
Gradient
  detection of, 78, 92, 98
  molecular, 67, 77, 89, 91, 97
  spatial, 72, 82, 84–85, 127, 129
Granovetter, M. S., 45
Granular, 79, 82, 109
Gray matter
  cortex organization, 79, 144
  cortical folding, 131–136, 138, 140
  evolution, 39
  pathological changes, 162, 166, 169, 174
Grossman, N., 188–189
Growth
  axon, 95–99
  duplication-divergence, 114–115
  nonlinear (accelerated), 113–118
  old-gets-richer, 112–113, 116, 148
  parallel, 122–123
  parcellation theory, 114
  preferential attachment, 111–112
  preferential detachment, 115, 148–150
  rich-gets-richer, 111
  serial, 123
  spatial, 31, 102–104
Guido, W., 180
Guimera, R., 18, 108
Güntürkün, O., 48
Gurney, K., 24, 109
Guy, J., 83
Gyrencephalic, 131, 134, 138, 140
Gyrification
  folding theories, 132–134
  models for, 137–140
  across primates, 131
  scaling laws, 138

Hagmann, P.
  on adolescence network changes, 129, 145, 150
  on ddiffusion tensor imaging, 5, 17, 48, 60
Hahn, J. D., 55
Haier, R. J., 151
Hallucinations, 33, 155, 165
Hatching, 44, 98, 112, 120–121
Headaches, 187
Hebbian learning, 75, 129
Hedderich, D. M., 134
Helfrich, R. F., 191
Hellwig, B., 95
Helmstaedter, M., 43
Hely, M. A., 165
Hemineglect, 192

Hennig, M. H., 75
Henriksen, S., 94
Hensch, T. K., 121
Hentschel, H. G. E., 89, 98
Herculano-Houzel, S., 5, 58–59, 86, 131–132, 138, 140
Hermann, B., 158
Hermaphrodite, 37, 43–44, 81
Heterochronicity, 127–129
Heterotopia, 110, 158
Heuer, K., 131
Hierarchy, 51, 93
Hilgetag, C. C.
  on connectomes, 26, 58–59, 95, 104, 115
  on cortical folding, 133–134
  on cytoarchitecture, 82
  on modular organization, 16, 25, 56, 150
  on small-world features, 24–25, 40
  on spatial growth, 31, 102–103
  on time windows and heterochronicity, 127–129
  on wiring optimization, 28–29
Hill, J., 145–146
Hindbrain, 82
Hinton, G. E., 167
Hippocampus
  brain diseases and, 148, 159, 169, 176
  brain stimulation of, 188, 190
  internal connectivity, 22, 50–51, 121
  layer organization, 79
His, W., 7
Holmes, A. J., 41
Homae, F., 129
Homeostasis, 78, 129, 148, 181
Homo sapiens. See Human
Homology, 55, 83
Hong, S. J., 110, 157
Hornykiewicz, O., 165
Horstmann, H., 7
Horton, J. C., 25
Horvitz, H. R., 44
Hoshooley, J. S., 48
Hosoda, C., 94

Hubel, D. H., 68–69
Huberman, A. D., 74
Hubs. *See also* Old-gets-richer; Preferential attachment; Preferential detachment; Rich-gets-richer
  brain diseases and, 153, 162, 167, 170–172, 185
  in connectomes, 12, 43, 48–49, 53, 55–56, 109–110, 145
  formation, 40, 107–118
  functional role, 46, 48, 60, 108–110, 148, 151, 174
  robustness, 23, 178–179
Human
  brain stimulation, 189–193
  connectivity, 5, 9, 17, 60, 63, 94
  cortical folding, 131–132, 137, 140
  evolution, 40–41, 83–85
  gene expression, 72–73
  hubs and rich-club, 110, 115, 145
  scale-free organization, 22–23
  small-world and modular organization, 40, 127–129, 145, 159–161
  spatial organization, 7, 26–27, 93
Human Connectome Project, 60, 136, 139, 144
Humphries, M. D., 24, 109
Hutchings, F., 194
Hutchinson, E., 158
Huttenlocher, P. R., 120, 144
Hyperexcitability, 158, 161, 197
Hypersynchronization, 196
Hypothalamus, 55

Illiterate, 147
Inflammation, 158
Ingalhalikar, M., 150
Inhibition
  brain stimulation and, 190, 192
  across fruit fly connectome, 47
  homeostasis, 182
  imbalance, 158
  lateral, 71–72, 74–77
  tonic, 157

Injury
  axonal sprouting, 180
  functional compensation, 177, 183–186
  functional connectivity changes, 181
  lesion timing, 186
  lesion types, 177
  recovery after lesions, 191
Innervation, 69–70, 72, 74, 92, 133
Innocenti, G. M., 93, 95
Insula, 109, 135, 156, 192
Intelligence, 150–151
Intestine, 38
Invertebrates, 39, 45
Isingrini, M, 151
Isocortex, 79–80
Isolated node, 13
Ito, M., 98
Iturria-Medina, Y., 19

Jaccard index, 32, 183
Jackson, G. M., 157
Jackson, S. R., 157
Jacobs, R. A., 127
Jardim-Messeder, D., 55
Jbabdi, S., 161
Jellyfish, 37
Jiang, L., 162
Jiang, X., 159
Jirsa, V., 77
Johansen-Berg, H., 161, 180–181
Jones, D. K., 143
Jones, D. T., 172
Jones, E., 60
Jucker, M., 166, 168
Jung, R. E., 151

Kahn, D. M., 114
Kaiser, M.
  on axon growth, 95–99
  on connectomes, 25, 38, 40, 45
  on hierarchical modular networks, 40, 150, 159–160
  on hub formation, 112–115, 148
  on network analysis, 13
  on network dynamics, 33
  on network growth, 86, 95, 108, 111
  on network robustness, 23, 170, 178–179
  on spatial features, 26–32, 59, 94
  on spatial growth, 102–104
  on time windows, 47, 120, 122–127
  on wiring optimization, 28–31
Kaliningrad, 8
Kalisman, N., 99
Karbowski, J., 29
Katz, L. C., 121
Katzman, R., 185
Kaufman, A., 38, 50, 92
Kaufmann, T., 162
Keck, T., 178
Kekic, M., 192
Kelava, I., 38
Keller, C. J., 190
Kelsch, W., 120
Kendler, K. S., 34
Kensinger, E. A., 137
Khambhati, A. N., 190
Kim, H. R., 173
Kim, J. S., 159–160
Klein, D., 134, 146
Koch, C., 147
Koene, R. A., 65
Kohonen, T. K., 77
König, P., 30
Königsberg bridge problem, 8
Koser, D. E., 92
Kötter, R., 20, 48, 58–59
Krubitzer, L., 114
Kuan, C.-Y., 81–82
Kubanek, J., 189
Kuhn, A. A., 194, 197
Kuramoto oscillator, 196

LaMantia, A. S., 69, 144
Larson, T. A., 48
Larva, 43–44
Lateralization, 31

Latora, V., 15–16, 149
Lattice, 21, 38, 40, 93, 153
Lau, Y. C., 157
Laughlin, S. B., 30
Lebel, C., 143
Leech, 43
Legon, W., 189
Leinenga, G., 189
Lemaire, T., 189
Lemieux, L., 160
Lesion studies. *See* Injury
Lewis, J. D., 156
Lewy body dementia, 165–167, 170
Li, A., 51
Li, H. J., 162
Li, L. M., 190, 192
Li, W. C., 97, 99
Lichtman, J. W., 7, 69, 102, 120
Lifestyle, 166
Lim, S.
  on preferential detachment, 148–150
  on time windows, 47, 122–124
Limbic cortex, 82, 120, 156, 167
Lin, H. Y., 157
Lin, M. K., 59, 71
Lissencephalic
  marmoset monkey, 59
  mouse, 50
  reelin gene, 83
  scaling-law, 138
Little, G. E., 92
Liu, Y., 155
Liu, Y. Y., 196
Lo, C.-Y., 169
Lohof, A. M., 69
Lomber, S. G., 183
Long, P., 94
Longitudinal studies, 63, 144, 151, 156–157, 159, 169, 174
Loops, 8
Lorenz, K. Z., 121
Lövdén, M., 166
Luders, E., 134–135

Luft, C. D., 196
Lund, J. S., 93
Luo, L., 95, 144

Macaque
  brain stimulation, 189
  connectivity, 19, 23, 26, 32, 40, 58–59
  hubs, 109, 112–113, 178
  network development, 83–86, 99, 102, 104, 115
  small-world organization, 24–25
  spatial network organization, 26–31, 93, 115, 134
Macroconnectome, 33, 51, 56, 60, 161
Madan, C. R., 137
Magnetoencephalography (MEG), 169, 181
Majka, P., 59, 61
Mallamaci, A., 71
Malnutrition, 162
Mammals, 14, 31, 37, 50, 59, 91
Mander, B. A., 148
Manger, P. R., 55
Marchiori, M., 15–16, 149
Markov, N. T., 58
Marmoset monkey, 43, 58–61, 131
Masuda, N., 29
Matching index, 183
Matthew's effect, 111
Maupas, E., 44
McKeith, I. G., 165
Mechanoreceptors, 44, 46, 92
Meinhardt, H., 71–72
Merzenich, M. M., 161
Mesencephalon, 39
Metabolic constraints, 28, 118, 131, 170, 178
Metazoans, 37
Metencephalon, 39
Metropolis, N., 28
Meunier, D., 150
Meynert, T., 79
Mezias, C., 173
Micheloyannis, S., 155
Microcircuit, 25, 72, 82, 155–156

Microconnectome, 33, 51, 56, 115
Microspheres, 9
Microstate, 170
Microtubules, 138
Midbrain, 167
Milgram, S., 24
Migration
 pathological changes, 155, 158
 radial, 67, 80–82, 86, 88, 133–134
Mikula, S., 51
Milham, M. P., 59
Miller, K. L., 60, 144
Milo, R., 19–20, 58
Minerbi, A., 148
Minicolumns, 195
Minimum spanning tree, 28
Minshew, N. J., 157
Misfolding proteins, 166, 185
Mishra, V., 148
Mitochondrial, 26, 191
Modules
 in connectomes, 45–46, 48–49, 51, 56, 58, 60
 evolutionary origin of, 38–40
 formation, 119–129, 145, 148–150
 hierarchical networks, 25, 56
 measures for, 16–19, 159–160
 pathological changes, 155–156, 158–160, 169
Mohades, S. G., 94
Mohan, U. R., 190
Mollusks, 39
Monod, J., 114
Mood, 158, 192
Morphogen, 70–72, 77, 104–105
Mosaic, 65, 71–72
Mota, B., 132, 138, 140
Motif, 19–20, 48, 58–59, 166
Motoneuron, 69, 72, 74
Mountcastle, V. B., 25
Mouse
 brain stimulation, 94
 connectome, 50–51
 development, 94, 121, 127–129
 layer architecture, 82–83, 85
 retina, 43
Muir, D. R., 93
Murray, J. D., 70–71, 91
Murre, J. M., 31
Mus musculus. See Mouse
Mustela putorius furo. See Ferret
Myelencephalon, 39
Myelin
 conduction speed, 12, 45, 91, 93
 early myelination, 110
 hub formation and, 118
 loss of, 169
 sheath establishment, 82, 91, 93–94, 143

Nathan Kline Institute (NKI), 139, 144, 175
Natu, V. S., 143
Nawa, H., 80
Nemathelminthes, 38, 43
Neocortex
 brain stimulation model, 194
 evolution, 83
 growth model, 84
 layer organization, 82
 module formation, 121
 old-gets-richer, 112–113
Neonates, 144, 148, 158
Nepusz, T., 33
Network flow, 185
Networks. *See* Connectivity
Neumann, W. J., 197
Neural crest, 67
Neurogenesis, 48, 84–85, 120, 123, 134, 145, 183
Neuromodulation, 187, 189, 193–194
Neuromorphic fields, 70–72
Neuromuscular junction, 44, 72
Neurotransmitter, 47, 79, 161, 174
NeuroVIISAS, 51
Neurulation, 67, 81
Newman, M. E., 16–17
Ng, Y. S., 191
Nguyen, J. A., 192

Nie, J., 132–133
Nikolic, K., 194
Nisbach, F., 127
Nitsche, M. A., 187, 190, 192
Noise, 29, 34, 188, 195
Non-metric multidimensional scaling (NMDS), 183–184
Nordahl, C. W., 135
Notch, 71
Nottebohm, F., 48

Ocelli, 39, 45, 47
Octopus, 37, 39
Oh, S. W., 37, 50
Okujeni, S., 129
Old-gets-richer, 112–113, 116, 148
O'Leary, D. D., 95, 144
Olfactory system
  dementia progression, 167, 173
  fasciculation, 98
  fruit fly, 44
  layer organization, 79
  module formation, 121
Oligodendrocyte, 91, 93–94, 169
Omnivores, 58
Optimization. See Component placement optimization (CPO)
Optogenetic stimulation, 94, 188–189, 194
Oscillations, 47, 188, 191–192, 194, 196–197
Overconnectivity, 145
Overrepresentation, 81
Overshoot, 183

Paleocortex, 79, 112–113
Palla, G., 16
Pallium, 83
Palmqvist, S., 165
Pandya, S., 176
Pang, T., 135
Parallel growth, 122–123
Parcellation, 5, 10, 16–17, 56, 71, 128
Parcellation theory, 114
Parkinson's disease
  as a type of dementia, 165
  diffusion progression model, 176
  focused ultrasound treatment, 189
  histological changes, 166
  limbic areas, 82
  models of deep-brain stimulation, 197
Participation coefficient, 18–19, 150–151, 159
Passingham, R., 58
Path length. See Characteristic path length
Pathoconnectomics, 153
Paul, L. K., 157
Payne, B. R., 183
Peraza, L. R., 174–176
Pereira, J. B., 169
Periallocortex, 80
Perilesion, 180, 185, 191
Perin, R., 119–120
Petanjek, Z., 144
Peters' Principle, 92
Pettersson-Yeo, W., 155–156, 161
Pharynx, 44, 81
Piano learning, 94
Picco, N., 84
Picker, A., 71
Pigeon, 37, 43, 48–49
Plaques, 58, 170
Plasticity
  activity- and spike-time dependent, 78, 129, 161, 181–183, 190
  and brain stimulation, 191, 194
  across development, 121, 144, 147, 179
  functional, 89, 161
  sleep and memory, 147
  structural, 89, 181–183
  synaptic, 78, 191, 194
Platyhelminthes, 38
Polania, R., 191
Polymicrogyria, 110, 158
Pons, 167
Positron emission tomography (PET), 169–171
Preferential attachment, 103, 111–112, 114, 116–118

Preferential detachment, 115–116, 148
Presubiculum, 80
Price, C. J., 33, 177
Price, D. J., 95
Primates, 40, 59, 84, 93, 119–121, 131, 180–181. *See also* Human; Marmoset monkey; Rhesus monkey; Squirrel monkey
Proisocortex, 80
Protein-protein interaction networks, 26, 110, 115
Protocerebrum, 39
Pruning
 of axons, 102, 144
 for fiber tracts, 129, 144–145
 and neurodevelopmental disorders, 161–162
 of synapses, 120, 144, 183
Puberty, 60, 144
Purkinje cells, 70, 89
Purves, D., 69, 102, 120
Putamen, 156
Putcha, G. V., 81

Qubbaj, M. R., 77

Radial-unit hypothesis, 132
Ragsdale, C. W., 83
Raj, A., 35, 172–173
Rakic, P.
 on cortical columns, 25, 65
 on critical periods, 31
 on pruning, 69, 144
 on radial-unit hypothesis, 132
 on time windows, 120
 on transient connectivity, 95
Rakic, S., 88
Ramón y Cajal, S., 7
Rapoport, J. L., 33
Rat
 brain stimulation, 190, 194
 connectivity, 26, 40, 51–55, 95
 filling fraction, 99
Rattus norvegicus. *See* Rat
Ravasz, E., 58, 114

Reato, D., 190
Recovery, 177, 179–181, 183–186, 191–192
Redundancy, 178
Reeler mouse, 82–83, 85
Rees, G., 157
Rehabilitation, 180, 191–192
Reichert, H., 81
Reijneveld, J. C., 153
Reinhart, R. M. G., 192
Rényi, A., 21, 116
Retina
 focal lesion, 181
 mechanosensing ganglion cells, 92
 mosaics, 65, 71
 ocular dominance, 121
 retinal waves, 75, 91
 topographic mapping, 68, 76–78
Retrograde tracers, 9, 52, 60–61
Rhesus monkey. *See* Macaque
Rhinencephalon, 79
Ribeiro, P., 20
Rich-gets-richer, 111
Riddle, D., 81
Riley, E. P., 80
Riley, K. P., 95
Robinson, P. A., 77
Robustness, 23–24, 30, 112, 178–179, 183–186
Rockel, A. J., 83
Rockland, K. S., 93
Rodent. *See* Mouse; Rat
Ronan, L., 133–135
Ropireddy, D., 51
Roundworm, 37–38, 43
Rubenstein, J. L., 161
Rubinov, M., 7, 10, 153, 183
Ryan, K., 43

Sabuncu, M. R., 151
Sadek, A. R., 51
Salami, M., 93
Sanchez-Rodriguez, L. M., 198
Scale-free networks, 22–23, 111

Scannell, J. W., 7, 29, 37, 40, 45, 56, 104, 184
Schizophrenia
    disconnection hypothesis, 161
    early-onset, 162
    genetic factors, 156
    increased gyrification, 135
    neurodevelopmental disease, 155
    pathological layer formation, 80
    rich-club organization, 110
Schlegel, A. A., 94
Schlemm, E., 157
Schmitt, O., 51
Scholz, J., 94
Schrepf, A., 158
Schumacher, J., 170
Schüz, A., 92, 102
Schwikowski, B., 26
Scorcioni, R., 90
Segev, R., 26
Seguin, C., 59
Sejnowski, T. J., 30, 167
Seldon, H. L., 132
Self-organization, 26, 34, 104
Self-organizing maps, 76–77, 185
Semaphorin, 92
Senden, M., 110
Serial growth, 123
Sernagor, E., 71
Seung, H. S., 7
Shanahan, M., 37, 48–49
Shaw, P., 120
Sherry, D. F., 48
Shih, C. T., 45–47
Shimizu, T., 48
Shortcuts, 21, 29, 40, 45, 149, 179
Silva, C. G., 65
Sinha, N., 161, 163
Skudlarski, P., 160
Small-world networks, 21, 24, 29, 45
Social networks, 8, 16
Socioeconomic data, 63, 150
Sole-Casals, J., 151
Song, H., 97

Sowell, E. R., 143
Spear, P. D., 180, 183
Specialization, 40, 45, 114
Sperry, R. W., 92, 95
Sporns, O.
    on connectomes, 7–8, 25, 60
    on degeneracy, 33
    on network analysis, 7, 10, 150, 159
    on network hubs, 47, 109
    on network motifs, 20, 48, 59
Squid, 91
Squirrel monkey, 58–59
Staiger, J. F., 83
Stam, C. J., 25, 129, 153, 167, 169
Stauffer, D., 8
Stein, D. G., 177, 186
Stepanyants, A., 99, 102
Stephan, K. E., 58, 155, 161
Sterratt, D., 72, 76, 147
Stief, F., 161
Stoll, E., 86
Stomatogastric ganglion, 38
Stoykova, A., 71
Striatum, 192, 197
Striedter, G. F., 29, 31, 37, 40, 132
Strogatz, S. H., 24, 29, 40, 45
Structural connectivity
    brain diseases and, 33–35, 110, 155–159, 169
    brain stimulation and, 190, 192, 196–197
    during development, 115, 145, 147, 149–151
    reconstruction, 8–11
    rewiring after lesions, 181, 184
    across species, 56, 60
    transient, 95, 144, 174
Stryker, M. P., 68
Stumpf, M. P. H., 23
Subcortical structures, 51, 94, 109–110, 143, 145, 156, 176
Subiculum, 50
Sukhinin, D. I., 55
Sulci, 109, 131, 134–135, 138, 173
Sulston, J. E., 44

Sun, Y., 156
Sundaram, S. K., 160
Sunkin, S. M., 51
Supekar, K., 129, 150
Superinnervation, 69, 74
Surgery, 158, 163, 177, 179, 186, 189
Swanson, L. W., 51, 53–55
Swindale, N. V., 74
Synapses
   establishment, 88, 120, 123, 144
   plasticity, 78, 191, 194
   pruning, 120, 144, 183
   tenacity, 148
Synchrony
   brain stimulation, 191–192, 196
   enhanced for depression, 158
   information integration, 29, 151
   module formation, 129
   role of hubs, 107

Tadpole, 43, 97, 99
Takahashi, E., 58–59, 61
Tallinen, T., 133, 138
Tamnes, C. K., 143
Tau proteins, 166, 170, 172
Tautochronicity, 127–128
Taylor, M. J., 157
Taylor, P. N., 158, 196
Tecchio, F., 191
Telencephalon, 5, 39, 48–49, 83
Thalamus
   connectivity between nuclei, 55
   in connectomes, 51, 92, 99, 170
   focused ultrasound stimulation, 189
Thanarajah, S. E., 156
Thiebaut de Schotten, M., 147
Thornton, C., 194
Tomasi, S., 93
Tomassy, G. S., 94
Tomsett, R. J., 195
Tononi, G., 33, 177
Topographic mapping, 68, 70, 76–78
Toro, R., 71, 131

Tourette syndrome, 157
Towlson, E. K., 112
Tracers, 59, 61
Transcranial alternating current stimulation
   (tACS), 188, 191–192
Transcranial direct current stimulation (tDCS),
   188, 191–192, 194
Transcranial magnetic stimulation (TMS),
   187–189, 192–193
Transcription, 73, 104
Transgenic animals, 59
Transient connectivity, 95, 144, 174
Trappenberg, T. P., 147
Tremor, 165, 189
Triplett, M. A., 76
Tschentscher, N., 151
Tuch, D. S., 7
Tumors, 177–178
Turing, A. M., 34, 70, 105
Tymofiyeva, O., 129, 145

UK Biobank, 60, 63, 144
Ultrasound, 188–189, 193
Unmyelinated fibers, 91, 93–94
Urbach, J. S., 92, 97
Utton, M. A., 172

Vaessen, M. J., 158
van den Heuvel, M. P.
   on brain development, 68, 110, 129, 145
   on comparative connectomics, 43
   on intelligence, 151
   on rich-club organization, 47, 56
   on schizophrenia, 160
Van Essen, D. C.
   on brain connectivity, 7, 37, 40, 58
   on brain surface shape, 7, 131, 136–137
   on Human Connectome Project, 60, 144
   on white matter pulling theory, 132–133
van Ooyen, A.
   on axon growth and fasciculation, 89, 92,
      98, 102
   on early pattern formation, 67

on structural plasticity and homeostasis, 181–183, 186, 191
on superinnervation, 69
van Rossum, M. C., 147
Varier, S., 40, 98, 112–113, 115, 120, 148, 186
Varshney, L. R., 115, 123
Váša, F., 145
Vazou, F., 151
Vázquez, A., 115
Ventricles, 80, 83, 132, 169, 173
Verhagen, L., 189
Vertebrates
  brain evolution, 39–40
  cell death, 81
  muscle fibers, 69
VERTEX, 194
Virus, 50, 188
Voets, N. L., 135
von der Malsburg, C., 30, 75–76
von Neumann, J., 30

Wagner, A., 178
Wang, Y.
  on cortical folding, 131, 133, 136, 139–140
  on neuromodulation, 193, 195
Wasserman, S., 8
Watanabe, T., 157
Watts, D. J., 15, 24, 26, 29, 40, 45
Weickenmeier, J., 173
Welker, W., 134
Wen, H., 157
Wen, Q., 28, 31
Wernicke, 154
White matter
  cortical folding, 131–136, 138
  evolution, 37–40
  organization, 25, 31, 92–94, 143, 145
  pathological changes, 154–155, 158, 169, 181
  prion spreading, 166, 173
White, J. G., 7, 37–38, 43–44, 98
White, T., 132
Widge, A. S., 192

Wiecki, T. V., 197
Wiesel, T. N., 68, 70, 93
Williams, D. L., 157
Williams syndrome, 135
Willshaw, D. J., 69, 71–72, 74–76
Wiring optimization, 28–31
Wolf, L., 50
Woodward, N. D., 155
Wörgötter, F., 89, 181–182, 191
Wulst, 83

Xie, L., 147
Xu, G., 133
Xu, Y., 145

Yamamoto, N., 7, 38, 92
Yan, G., 196
Yang, Z., 162
Yeast, 26, 178
Yeh, F. C., 94
Yizhar, O., 161
Yoo, S. W., 185
Yoshiyama, Y., 171
Young, M. P., 7, 16, 29, 37, 40, 51, 58, 104, 119
Ypma, R. J., 50
Yu, Y. C., 121

Zalesky, A., 160
Zamora-Lopez, G., 48
Zecevic, N., 88
Zhan, Y., 161
Zhang, Z., 158
Zhao, T., 145, 148, 150
Zhou, C., 56, 110, 161
Zhu, F., 51
Zielinski, B. A., 156
Zingg, B., 50, 52
Ziv, N. E., 148
Zubler, F., 86–87, 93, 104–105
Zuo, X. N., 143